AGAINST ECOLOGICAL SOVEREIGNTY

posthumanities

CARY WOLFE, SERIES EDITOR

AGAINST ECOLOGICAL SOVEREIGNTY

Ethics, Biopolitics, and Saving the Natural World

MICK SMITH

posthumanities 16

 University of Minnesota Press
Minneapolis
London

An earlier version of chapter 2 appeared as "Worldly (In)difference and Ecological Ethics: Iris Murdoch and Emmanuel Levinas' Environmental Ethics," *Environmental Ethics* 29 (2007): 23–41; reprinted with permission of *Environmental Ethics*. An earlier version of chapter 3 appeared as "Wild Life: Anarchy, Ecology, and Ethics," *Environmental Politics* 16 (2007): 470–87; reprinted with permission from Taylor and Francis, http://www.informaworld.com. An earlier version of chapter 4 appeared as "Citizens, Denizens, and the Res Publica: Environmental Ethics, Structures of Feeling, and Political Expression," *Environmental Values* 14 (2005): 145–62; reprinted with permission of the White Horse Press; and as "Suspended Animation: Radical Ecology, Sovereign Powers, and Saving the Natural World," *Journal for the Study of Radicalism* 2, no. 1 (2008): 1–25; reprinted with permission of Michigan State University Press. Portions of chapters 5 and 6 were previously published as "Ecological Citizenship and Ethical Responsibility: Arendt, Benjamin, and Political Activism," *Environments Journal* 33 (2005): 51–64; reprinted with permission of *Environments Journal;* and as "Environmental Risks and Ethical Responsibilities: Arendt, Beck, and the Politics of Acting into Nature," *Environmental Ethics* 28 (2006): 227–46; reprinted with permission of *Environmental Ethics*. An earlier version of chapter 7 appeared as "Against Ecological Sovereignty: Agamben, Politics, and Globalization," *Environmental Politics* 18 (2009): 99–116; reprinted with permission from Taylor and Francis, http://www.informaworld.com.

Published by the University of Minnesota Press
111 Third Avenue South, Suite 290
Minneapolis, MN 55401-2520
http://www.upress.umn.edu

Library of Congress Cataloging-in-Publication Data

Smith, Mick.
    Against ecological sovereignty : ethics, biopolitics, and saving the natural world / Mick Smith.
        p.   cm. — (Posthumanities 16)
    Includes bibliographical references and index.
    ISBN 978-0-8166-7028-4 (hc : alk. paper)
    ISBN 978-0-8166-7029-1 (pb : alk. paper)
    1. Environmental ethics.   2. Human ecology—Philosophy.   3. Biopolitics—Philosophy.   4. Sovereignty—Moral and ethical aspects.   5. Culture and globalization.   I. Title.
    GE42.S575 2011
    179'.1—dc22
                                                                                2011011173

Printed in the United States of America on acid-free paper

The University of Minnesota is an equal-opportunity educator and employer.

18 17 16 15 14 13 12 11   10 9 8 7 6 5 4 3 2 1

For my father,
Alan Smith
1931–2009

CONTENTS

ACKNOWLEDGMENTS

The world is a wonderful place despite the ecological damage we inflict on it. Our existence in the world is finite, but of all the experiences it offers, love is that which matters most. My father was, quite simply, the loveliest man I have ever known, and I miss him deeply. I dedicate this book to him and also to my mother, Eileen, my partner, Joyce Davidson, and Emily Rain, our amazing daughter.

Because this book concerns the more-than-human world, it is only fitting that it should also be dedicated to Sammy, a cat whose company over many years mattered more than I could ever say, and to Furdie, Horace, and Tiki, who all contributed in their own singular ways.

It would be impossible to mention all those whose ideas inspired this work, not least by their ecological activism. I thank Sue Donaldson, Victoria Henderson, and Heather Schmitt for reading and commenting on the entire manuscript and Cate Mortimer-Sandilands for inviting me to present an early version of "Suspended Animation: Radical Ecology, Sovereign Powers, and Saving the Natural World" to the Nature Matters conference she organized with Meghan Salhus at York University, Toronto, in 2008. Cate also gave me thoughtful comments on later parts of this manuscript. I thank the two anonymous readers for the University of Minnesota Press and Cary Wolfe, Posthumanities series editor, for extensive and supportive comments. Thanks also to Douglas Armato and Danielle Kasprzak at the University of Minnesota Press. As usual, the errors remaining are my own.

INTRODUCTION

A Grain of Sand

> It is a contest, and a grain of sand turns the balance.
>
> —Charles Darwin

DARWIN'S REMARK REFERS, of course, to evolution, the understanding of which is certainly not without political connotations. But what radical ecology contests is human dominion over the natural world—that is to say, *ecological sovereignty* in all its many guises. This political contest too may turn on a grain of sand, on a few words, deeds, or circumstances that might alter the pattern of that future predicted by (and predicated on) today's ecologically and socially destructive forms of life.

This contest is political because human dominion over the Earth is not, as so many assume, just a theological idea(l) justified by biblical exegesis or a secular ideology unquestioningly assumed by (supposedly self-critical) Western philosophical systems. It is also the key principle, both theoretically and practically, that underlies the modern political constitution. Here, *modern constitution* should be understood both in Bruno Latour's (1993; 2004, 239) "broader metaphysical sense," as the explicit (but never fully achievable) modernist division of the world into two realms—the human and the nonhuman, subjects and objects, evaluatively driven politics and the supposedly apolitical, value-free, natural sciences, and so on—and constitutionally in the narrower political sense: the modern principle of national sovereignty, for example, presumes ecological sovereignty over a specific territory (Kuehls 1996).

Ecologically speaking, competing claims to territorial sovereignty, such as those concerning an Arctic seabed now increasingly bereft of its protective ice cap, are all about which state gets to decide how and when these "natural resources" are exploited. Of course, states may also employ ecological rhetoric in staking their claims to be responsible stewards of nature. But making such decisions, even if they occasionally involve distinguishing between natural resources and nature reserves, is the defining mark of ecological sovereignty, and these decisions are premised on, and expressions of, the modernist metaphysical distinction between the decisionistic politics associated with (at least some) "properly human *subjects*" and the objectification of nonhuman nature as a resource. The modern constitution and its overseer, the principle of ecological sovereignty, exemplify what Agamben (2004) refers to as the "anthropological machine"—the historically variable but constantly recurring manufacture of metaphysical distinctions to separate and elevate the properly human from the less-than-fully-human and the natural world.

Contesting ecological sovereignty requires that we trace connections between such metaphysical distinctions and political decisions. It requires (to employ a somewhat hackneyed phrase) yet another Copernican revolution—a decentering, weakening, and overturning of the idea/ideology of human exceptionalism. We might say that any critique of political sovereignty failing to attend to these metaphysical distinctions will be ecologically blind, whereas any ecological critique of humanist metaphysics in political isolation will be empty. For example, past environmental critiques of human dominion and debates about the merits of Earthly stewardship (White 1967; Black 1970; Passmore 1974) may have been vital catalysts for the emergence of radical ecology, but they rarely touched the principle of sovereignty itself, still less recognized its political ramifications. Yet if we keep the political principle of sovereignty intact, then we automatically and continually give shelter to the notion of ecological sovereignty, and all talk of changed ecological relations is ultimately hollow.

Of course, few ecologists are going to protest if a sovereign nation decides to set aside an area as a nature reserve! But the point is that this decision, which divides and rules the world for ostensibly different purposes, is plausible only if the overarching authority to make (and adapt and reverse) such all-encompassing decisions is already presumed. It presumes human dominion and assumes that the natural world is already, before any decision is even made, fundamentally a human resource. This is, after all, both the contemporary condition that nature is being reserved (and yet not released) *from,* and the original condition of that mythic prepolitical "state of nature" (epitomized in Locke's work) where a presumptive ecological sovereignty serves as the foundational premise for an emergent political sovereignty (see chapter 3). How paradoxical, then, that the decision to (p)reserve some aspects of ecology, to maintain it in what is deemed to be its natural state, has today become a matter of political sovereignty. Paradoxical because, without all nature being initially assumed to be a resource, there would be no original justification for political sovereignty: And yet, without political sovereignty, so the story now goes, nature cannot be preserved from being treated as a resource.

Either way, one might say, everywhere *sovereignty* declares nature free, it is already in chains. And metaphysically, ecologically, and politically speaking, the claims and chains of sovereignty are all-encompassing: they encircle the world. In this sense, sovereignty is an antiecological and not, as its accompanying rhetoric and its modern environmental proponents (see chapter 7) sometimes suggest, a potentially ecological principle—at least if we understand ecology as something more than, and irreducible to, a human resource, and this is radical ecology's (but certainly not only radical ecology's) understanding.

Another way of putting this, and one that fits with the analysis of sovereignty provided by thinkers as politically diverse as Carl Schmitt, Walter Benjamin, and Giorgio Agamben, is to say that the nature reserve is the exception that decisively proves the rule—in the sense of both making tangible the dominant ideological norm and exemplifying the overarching principle and power of the ruling authority to decide. The nature reserve is exempted from being a resource, freed from human domination, only by being already and always *included* within the remit of human domination. And according to Agamben (2004, 37), this troubling figure of exclusion/inclusion, this "zone of indeterminacy," typifies the operation of both sovereignty and the anthropological machine.

Of course, to the extent that Schmitt, Benjamin, and even Agamben (see chapter 4) focus on political (and juridical) considerations, they tend to remain blind to sovereignty's metaphysical and ecological implications (although, as this book argues, Agamben's detailed discussion of the metaphysics of the anthropological machine in *The Open: Man and Animal* (2004) opens just such an unfulfilled possibility. These theorists are concerned with defending (in Schmitt's case) or critiquing (in

Benjamin and Agamben's) sovereignty's role as an *anti*political (and not an antiecological) principle. But any ecological politics obviously has to concern itself with critiquing both aspects of sovereignty. The paradox here, and one that mirrors claims to ecological sovereignty, is that in terms of modernist political mythology, sovereign power is derived from political activity, from, for example, participation in an original social contract. Yet Schmitt's *Political Theology* (2005, 5) opens with his famous definition, "Sovereign is he who decides on the exception"—that is, it is the ultimate mark of sovereign power to be able to suspend the normal rule of law and the political order by declaring a state of emergency (exception). The mark of sovereign power is precisely its exclusive ability to suspend political activities by virtue of its claim to have already subsumed (to inclusively represent) within itself all political activity—to declare an antipolitical condition, a state of emergency that abrogates the very activities that constitute its purported legitimacy.

Despite their political differences, Schmitt, Benjamin, and Agamben all agree that this definition of sovereignty holds no matter how democratic the modern state's political constitution may claim to be, although Schmitt, of course, infamously defended the legality of the state of emergency declared following Hitler's accession to power. This declaration took the form of the Decree for the Protection of the People and the State, which used emergency provisions *in* the Weimar constitution (Article 48) to suspend the personal liberties supposedly guaranteed *by* the Weimar constitution (Agamben 2005, 14–15). And since, as Agamben (1998, 28) remarks, "The decree was never repealed . . . from a juridical standpoint the entire Third Reich can be considered a state of emergency that lasted twelve years."

Sovereignty inevitably harbors such dangers because, although the justification for declaring a state of emergency is *always* the defense of the people and the state, sovereign power, by definition, takes it upon itself to decide what constitutes a danger, the state, and the people (and here again, in this last instance, the anthropological machine can play a devastating role, as it did in Nazi Germany, in deciding who does and does not count as properly human). Only politics *as such* can contest such decisions, but politics is precisely what is suspended as a consequence of this antipolitical decision. The sovereign decision deprives particular people within a particular territory of their right to engage in politics. Here, people's "political life" *(bios politikos)*, their capacity to express themselves individually and in community with others through their public words and deeds (and this is Arendt's [1958] definition of politics) is (politically) stripped from them. Their political existence is denied by a decision that reduces them to the condition Agamben refers to as "bare life" (naked existence). Bare life is not a return to a prepolitical state of nature, not a condition prior to or exceeding political authority, but a condition in which those included within the polity (subject to sovereign political authority) are nonetheless excluded from the polity (in terms of political participation and the protection supposedly afforded by its now suspended laws). They experience the worst of both worlds, being subject to a power under which they have no legitimate standing.

The prisoners in Guantánamo Bay and the residents of New Orleans following Hurricane Katrina both found themselves in this indeterminate (excluded/included) condition (Butler 2004; Gregory 2006), and Agamben contentiously argues that the most extreme example of such a zone of indeterminacy was the concentration camp. But as Benjamin (2006, 392) remarked, the "tradition of the oppressed teaches us that the 'state of emergency' in which we live is not the exception but the rule." In other words, the state of emergency is not confined to such extreme times and places but also becomes normalized in more diffuse ways.

To put this in Agamben's terms, which draws on Foucault's (2004; 2008) earlier analyses, we are now in a situation in which politics is everywhere being replaced by biopolitics, the governmental management and control of the biological life (and death) of populations. Even in the most democratic of countries, we find ourselves increasingly reduced for the purposes of governance to so much biological information, to the collection and manipulation of statistics (state information) concerning every aspect of our lives: from our birth date and height to our DNA profile, the unique patterns of our iris and fingerprints, the information in our biometric passports, and so on. The consequences of this more subtle but pervasive biopolitical way of reducing people to bare life are only beginning to be recognized because, as Esposito (2008, 44) argues, "The category of biopolitics seems to demand a new horizon of meaning, a different interpretative key."

This may be so, but the implications of biopolitics for ecology and the ecological implications of biopolitics have hardly even been noticed, let alone interpreted, for example, in terms of the parallels between the biopolitical reduction of people to bare life and the biopolitical reduction of more-than-human nature to resource, to "standing reserve" in Heidegger's terminology (see chapter 4). And if Hurricane Katrina exemplified an antipolitical response to an ecological crisis, what exactly does *ecological crisis* mean? This state of emergency occupied a zone of indeterminacy in more than one way, for, like many (perhaps all) modern ecological crises, it seems impossible to specify whether its causes were anthropic (in this case, human-induced global warming and inadequately engineered levees) or natural. Katrina itself is then an example of the ecological erosion of the basis of the modern constitution (see chapters 4 and 5). Today, only religious fundamentalists and insurance companies still claim to know what constitutes an act of God and/or of nature, and both, unsurprisingly, see them everywhere.

And what of global warming and the melting ice caps, the increased severity and unpredictability of storms, the effects of climatic changes on crops and peoples, the loss of topsoil and the unavailability of fresh water, the global pandemics threatened by new disease strains, the destruction of the rain forests, the acidification of the air and oceans, the extinction of so many of our fellow species, and on and on? Do these conditions not also constitute ecological crises, exactly as environmentalists and radical ecologists have been saying for so many years? Has so much of environmental politics not focused precisely on trying to get governments to take these impending crises seriously?

But now a new thought arises. What if sovereign powers take it upon themselves to decide that there is, after all, an ecological threat to people and state sufficient to warrant the definition "crisis"? *Isn't there now a real, and devastatingly ironic, possibility that the idea of an ecological crisis, so long and so vehemently denied by every state, will find itself recuperated, by the very powers implicated in bringing that crisis about, as the latest and most comprehensive justification for a political state of emergency,* a condition that serves to insulate those same powers against all political and ethical critique concerning their previous (in)activity and their (final) solutions? We might find that the global war on terror will segue seamlessly into the crisis of global warming, a condition produced by interventions in the natural world of a kind that were initially deemed politically unchallengeable by everyone *except* radical ecologists. And the political and ecological danger is that this emergency would be used to legitimate yet further technocratic interventions, to further extend the state and corporate management of biological life, including the continuing reduction of humanity to bare life and nature to mere resource, and to stifle ecological politics as such.

Anyone doubting such possibilities should familiarize themselves with the growing literature on environmental security (for critical appraisals, see Barnett 2001; Dalby 2002) or read works like those of respected scientist James Lovelock (see chapter 6). Lovelock, for so long a doyen of environmentalism, claims we are now on a war footing with an externalized but strangely personified nature, Gaia, our heartless "Earthly enemy" (Lovelock 2006, 153). (In making nature the personalized enemy without, he inverts the more usual *de*personalization of those sovereignty declares the political "enemy within.") The impending crisis of global warming might, he argues, necessitate the decisionistic suppression of certain political liberties: "rationing," "restrictions," a "call to service," and our suffering "for a while a loss of freedom" (153)—in the name of survival and security. This, he combines with a call to impose, without further political prevarication, high-tech solutions (for example, giant space-mounted sunshades to deflect incoming light, a massive expansion of nuclear power, and so on) based on advice provided to sovereign powers by a "small permanent group of strategists" (153). The solution, then, seems to be a more extreme (emergency) version of the biopolitical same!

For Lovelock, "There is no alternative" (153). But what politics as such suggests is that there are always alternatives, it is just that some of these alternatives may not be so palatable to those currently wielding authority and "making an economic killing." Radical ecology is all about providing a place to voice possible alternatives, to question, critique, and innovate. In particular, it challenges the view that there is no alternative to the growth-oriented capitalism that currently powers greenhouse gas production, species extinction, habitat destruction, and so on. In other words, it voices the kind of socioeconomic and political possibilities Lovelock does not even think worth contemplating. In opposition to a biopolitics that reduces the more-than-human world to material for resource management and political beings to a matter of bare life, it suggests a *provisional and constitutive ecological politics as such* (see especially chapters 6 and 7). This, as already argued, necessitates a political and ecological critique of the principle of sovereignty per se in all its forms. This must include any temptation by environmentalists to champion the "sovereignty of nature," the idea that nature itself should be what decides our politics (see chapter 4) or even to espouse, as Iris Murdoch (1970) does, the *Sovereignty of Good*—of ethics itself (see chapter 2). The breadth and depth of this critique is why radical ecology is potentially the most radical form of politics, why it offers the most fundamental challenge to the established (metaphysical and political) order of things.

Insofar as it contests the overarching metaphysical and political principle of sovereignty with all its exceptional and biopolitical connotations, radical ecological politics is *anarchic* (and hence its relation to anarchism and varieties of ecological anarchy requires elucidation, see chapter 3). It rejects the inversion of reality that defines politics as membership of a political citizenry always beholden to sovereign constitutional principles, emphasizing instead the creative mutualistic potential of politics as such. In place of the political paradigm of (human) citizenship, it suggests a constitutive ecological politics of subtle involvements and relations between more-than-just-human beings, the denizens who together compose the world in something like an ecological variant of Jean-Luc Nancy's sense (see chapter 7). The capacities to be so involved owe nothing to sovereign powers but emerge as features of ecology and politics as such. That which sustains this ecological politics, that lets politics, like ecology, be much more than just a contest, is *ethics*. An ecological politics would be impossible without an ecological ethics.

Here ethics is not to be understood as moralizing, following rules, the employment of a felicific calculus, or even the recognition of (more-than-human) rights, but in a more primordial sense—as what Levinas (1991, 304) too calls *anarchic* "first philosophy" (see chapter 2). From Levinas's perspective, ethics concerns those responsibilities that arise from encounters with Others *revealed in their singularity* as so very different from, and so irreducible to, that which our self-interested, possessive, and instrumental preoccupations would make of them. Ethics is first philosophy because it appears unbidden even before the possibility of self-reflection and deliberate self-interest emerges. It is anarchic because, like the unfathomable singularity of Others, it cannot be captured and contained under preconceived formulae, categories, or overarching principles (the most decisive of which is sovereignty).

Ethics takes responsibility for the effects of our actions on significant Others in the face of ineradicable uncertainties, including uncertainty about human subjectivity itself and the objective needs of

Others. Ethics and certainty cannot coexist and yet, even in the face of inevitable quandaries, ethics still insists on our being involved, on the impossibility of abrogating ourselves from responsibility even (indeed, especially) where these uncertainties might concern matters of life and death (see chapter 1). Ethically speaking, we cannot *not* be responsible for our actions. Ethical uncertainties also disturb all attempts to define once and for all to whom such responsibilities should extend—who counts, or does not count, as significant. Therefore, an ecological ethics necessarily challenges the metaphysical certainties produced by the anthropological machine in its attempts to limit ethical consideration to those classified as properly human (see chapter 2). This would have to include Levinas's own metaphysics, which are no less ecologically blind (Llewelyn 1991a; Wood 1999; Smith 2007a), as Derrida (1992, 278), referring to Heidegger's attempt to limit ethics to *Dasein* (effectively his own definition of the properly human being) but also to Levinas's remarks:

> Let us venture, in this logic, a few questions. For example, does the animal hear the call that originates responsibility? Does it question? Moreover can the call heard by *Dasein* come originally from the animal? Is there an advent of the animal? Can the voice of a friend be that of an animal? Is friendship possible for the animal or between animals? Like Aristotle, Heidegger would say: no. Do we have a responsibility to the living in general? The answer is still "no," and this may be because the question is formed, asked in such a way that the answer must necessarily be "no" according to the whole canonized or hegemonic discourse of Western metaphysics or religions, including the most original forms that this discourse might assume today, for example, in Heidegger or Levinas.

An ecological ethics awakens us to the wider more-than-human world. It goes much further than even Derrida suggests, raising questions concerning the singular significance of beings other than animals, too: trees, fungi, rivers, rocks.

As Derrida's passage reminds us, rejecting metaphysical humanism cannot be simply a matter of extending the role and rule of human subjectivity to include selected nonhuman species under the auspices of an ethics that remains the sovereign "property" of human subjects. Which also means that to escape the anthropological machine, an ethical concern for nature, and the politics associated with it, would need to be an expression of a relation not predicated on whether or not such a concern is properly human. Some beings other than humans also have ethical possibilities (Masson and McCarthy 1995; Bekoff and Pierce 2009).

Some of those classified as Homo sapiens seem more or less devoid of ethical feeling. Again, this illustrates the importance of understanding the multifarious ways in which the Earth's singular and irreducible denizens compose the world in contestation *and* in mutuality.

These, then, are some of the arguments presented here. But what follows makes no pretence to provide a history or even a genealogy of the concept of (ecological) sovereignty but to expose its antithetical relations to ethics and politics as such. Instead, it reconfigures a constellation of philosophical ideas and theories that, imagined in a different way, might add momentum to turn the ecological and political balance. It is not a blueprint for a future society, still less a book on environmental policy (see the apologue), but a call to take ethical and political responsibility for saving the diversity of a world whose "endless forms most beautiful and most wonderful" (Darwin 1884, 429) were created by, and through, the unpredictable and unrepeatable pathways of evolution. It argues that sustaining a place for ecological politics and saving the natural world both depend on rejecting the antipolitical and antiecological principle of sovereignty.

Scene of the Dead Man. Paleolithic cave painting at Lascaux, ca. 15,000 BC.
Copyright The Gallery Collection/Corbis.

1 AWAKENING

> It's not that what is past casts its light on what is present, or what is
> present its light on what is past; rather, image is that wherein what
> has been comes together in a flash with the now to form a constel-
> lation. In other words, image is dialectics at a standstill. For while
> the relation of the present to the past is a purely temporal, continu-
> ous one, the relation of what-has-been to the now is dialectical: is
> not progression but image, suddenly emergent.
>
> —Walter Benjamin, *The Arcades Project*

AMONG THE HUNDREDS OF IMAGES on the walls of the Lascaux
cave, mostly of horses and aurochs, but also including stags, ibex, and
bison, only one depicts a human figure. For Georges Bataille (2005),
this figure, with its bird head and its animal associate, located in the
deep shaft of the cave's apse, its "Holiest of Holies"[1] held a special
importance. The gutted bison, naturalistically rendered, pierced by a
spear and with its intestines unraveling, faces this apparently dead
but ithyphallic man. Unlike the bison and the cave's many other animal representations, all so skillfully depicted, this human figure is only
roughly sketched, but his cultural importance, then and now, is no less
significant for that. For Bataille, this image, above all others in Lascaux,
encapsulates the moment of humanity's inception, of our becoming rec-
ognizably human.

This is not at all to say that this "most ancient art" (Bataille 2005, 103)
is empirically the earliest example of what might be taken as a defining
capacity to produce representational art. (The Lascaux paintings, composed over an extended period between fifteen and seventeen thousand
years ago, are in any case now deemed relatively recent in comparison

with, for example, those discovered at Chauvet in 1994.) Rather, for
Bataille, this particular artistic representation provides an exemplary
expression and illustration of a specifically human existence, one that
(supposedly) unlike any other animal has become self-aware concerning the nature of its own mortality. This image reveals the presence of
human beings able to represent to themselves, in thought and in art,
both the inevitability of their own future deaths (the sprawled figure)
and the dependence of their ephemeral lives on occasioning the death
of other animals (the speared bison). What humanity discovered in its
relations with the animal world, says Bataille (2005, 173), is the "fact of
being suspended, hung over the abyss of death, yet full of virile force."

Yet we should be careful here not to read dominant modern sensibili-
ties into this situation. The human figure's erect phallus is not, Bataille
claims, indicative of the kind of virility all too often characterized by
that macho, warlike individualism that sets the archetypically masculine hunter above the animals he kills. His supine position confirms
Bataille's claim. Rather, as the painstaking care accorded to Lascaux's
images of living creatures—and they almost all depict living and not
dead or dying creatures—also attests, this image represents the transi-
tory vitality of human and animal lives and deaths, together with the
recognition of human responsibility for the deadly consequences that
the fulfillment of their desires has for other living beings.

According to Bataille (2005, 171–72, my emphasis),

> The man is *guilty* of the bison's death because a line coming from an ex-
> pressly drawn propellant penetrates the animal's stomach. Because the
> man is guilty, his death could therefore be taken as compensation offered by
> chance or perhaps voluntarily to the first victim. It is of course very difficult
> to assess with any accuracy the Palaeolithic artist's intention, but the *mur-
> der* of an animal required expiation from its author. The author had to have
> rejected that which weighed down on him from the killing of his victim: he
> himself fell from this act, prey to the power of death, and had to have at least
> purified himself of a marked stain. The representation in the pit grants this
> situation a final consequence: he who gives death enters into death.

The shared desire to live—the erect phallus, the still standing bison—
marks an affinity with the hunted animals, not an ontological separation. Indeed this therianthropic figure with bird head, together with
what appears to be a staff topped with a similar bird (possibly representing black grouse), also suggests the kind of affinity that typically
underlies shamanic transformations, both from human to animal and

from the realm of life to death. This bird, says Bataille (2005, 172), "signifies the shaman's voyage into the beyond, into the kingdom of death."[2] This affinity is also evidenced in the practices of sympathetic magic underlying those artistic acts that made manifest the desired animals among the already present, and hence not pictorially represented (Bataille 2005, 50), humans in the cave.

Let us imagine, says Bataille (2005, 51),

> before the hunt, on which life and death will depend, the ritual: an attentively executed drawing, extraordinarily true to life, though seen in the flickering light of lamps, [which surely adds movement and vitality to the drawings] completed in a short time, the ritual, the drawing that provokes the apparition of the bison. This sudden creation had to have produced in the impassioned minds of the hunters an intense feeling of the proximity of the inaccessible monster, a feeling . . . of profound harmony. . . . As if men, obscurely and suddenly had the power to make the animal, though essentially out of range, respond to the extreme intensity of their desire.

This creative act is magical and sympathetic both in the sense of the animals being drawn (represented and made to appear before those present) approvingly and beautifully lifelike, and in their subsequently being drawn *toward* the presence of the hunt that is to follow. Of course, the "sympathy" of sympathetic magic does not extend to not killing the animal, but nonetheless, as the image in the shaft depicts, the human desire to live through the hunt is also realized to entail a much more ambiguous desire for the death of another animal. In other words, this image seems to illustrate an ethical quandary coeval with the emerging awareness of the human being's own mortality whereby the temporary postponement of that inevitable human death brings about the permanent loss of Others' lives. On Bataille's reading, then, this painting at Lascaux is also indicative of the birth of humanity's ethical as well as artistic sensibilities, and what is more, of an ethical sensibility directed precisely toward nonhuman animals.

Of course, more needs to be said about the various ways this ethical sensibility might be composed; about the phenomenology and ecology of sympathies; about the (dis)associations between guilt, responsibility, and recompense. That Bataille chooses to emphasize the "guilty pleasures" associated with animal deaths is hardly surprising given his famously morbid obsessions. But his emphasis in no way detracts from how such a suggestion confounds human-centered and Hobbesian conceptions of a human nature (or of that mythical "state of nature" where such a nature is expressed without a cultural context; see chapter 3) wherein the individuals' self-recognition of their mortality is entirely divorced from any ethical concern for others, especially those animal others who are always already assumed to be no more than useful objects at our beck and call. Far from representing a claim of dominion over nature, the caves at Lascaux testify to an emergent humanity that, from its very inception, is marked by awareness of and a potential sense of responsibility for its creative/destructive worldly actions. "From the depths of this fascinating cave, the *anonymous, effaced* artists of Lascaux invite us to remember a time when human beings only wanted *superiority* over death" (Bataille 2005, 173; my emphasis). Of course, Bataille was also aware that this particular, and necessarily unrequited, desire flies in the face of the same self-understanding of human mortality that he, like Heidegger, regards as distinguishing human from animal existence.[3] For this reason, we might say that Lascaux suggests a troubled encounter with death but also a profound, though necessarily equivocal, celebration of life, human *and* animal.

Even accepting the tenor of Bataille's interpretation, we should be careful to specify what these images portend, which is not the hitherto unsuspected existence of an alternative human nature—in the sense of a biologically predetermined propensity to concern ourselves ethically about the lives of other animals. Rather, it shows that from humanity's very beginning, *which is impossible to pinpoint,* our self-understandings have been caught up with our troubled relations with the lives of other animals. The image in the shaft is exemplary not in the sense of providing a (forgotten) template for, or throwing light on, the constitution of all subsequent human existence, but in the sense that it comprises a "moment," a torsional effect, that twists our currently accepted sociocultural trajectory, revealing hitherto unsuspected ethical possibilities opened by the self-understanding of human mortality. As such, it is a moment exempted from, momentarily taken outside of, what Agamben (2004), following Jesi, calls the "anthropological machine," that is, the sociohistorically variable ways in which humanity has constantly redefined itself (metaphysically) over and against animality, as speaking animal, rational animal, enspirited animal, tool-using animal, and so on. What Lascaux now awakens is an understanding that the self-forming, constitutive powers of human communities need not be (as philosophy so often is) used to erect supposedly absolute definitional boundaries, or entail the diminishment of the lives (and deaths) of nonhuman others. Hence, the Lascaux image now offers the possibility of an unexpected twist *(Verwindung)* in the plots of the anthropological machine.[4]

Lascaux and the Anthropological Machine

> Every beginning supposes what preceded it, but at one point night gave birth to day and the daylight we find at Lascaux illumines the morning of our species. It is the man who dwelt in this cave of whom for the first time and with certainty we may finally say: he produces works of art; he is of our sort.
>
> —Georges Bataille, *Prehistoric Painting: Lascaux or the Birth of Art*

Bataille claims that Lascaux evidences a kind of prehistoric awakening concerning the profound mysteries of life, death, and desire, one that finds expression in the coeval emergence of human self-awareness, art, religion, and ethics. Of course, as Arthur Danto remarks, it would "never have occurred to the painters of Lascaux that they were producing *art* on those walls" (quoted in Bruns 1997, 33). We need a particular "historically effected consciousness" to interpret the overlapping associations and images in this way. But somehow, in ways we cannot fully grasp, the cave images drawn and given movement by the flickering *human-produced* firelight at Lascaux were supposed to draw forth those self-same animals in the *natural* light of day. The animals' *death* in the daylight world was evoked beforehand by their images as *living* animals within the dark, cold enclosure of the cave. Their living form is thereby transposed from a terrestrial to another, subterranean, world, though these communicating worlds were probably not envisaged as entirely separate but as entangled, twisted together, in mysterious ways. The animals' subsequent deaths also ensured the continuing life of this emerging human culture, a culture trying to waken to its own situation, to understand its own existence.

This, though, if Bataille is right, was a world as yet without human dominion, one that did not presume or need to argue the purported naturalness of the distinction (the setting apart and setting above) accorded by the human to the human, a distinction whose forms the anthropological machine would prove so culturally imaginative in producing. Nonetheless, to the degree that we, like Bataille, are tempted to regard the cave as exemplifying and illuminating only an instant in a historical continuum (a temptation Benjamin refers to as "historicism"),[5] an inexorable tendency leading toward the now dominant view of a humanity set apart from and above creation, the cave also is likely to be interpreted as, in Agamben's (2004, 37) terms, an incipient "zone of indeterminacy," a "state of exception." That is, despite the subtlety of his interpretations, Bataille often falls prey to thinking of the people of Lascaux as just a primitive precursor of modern society's conception of humanity, as the first of "our" sort. But the dangers here are twofold. First, even as we are surprised that the cave's inhabitants "confronted the animal not as though he were confronting an inferior being or thing, a negligible reality" (Bataille 2005, 49), such ethical possibilities slip away from us as they are reduced to an instant in a past that has already been irredeemably and irretrievably superseded. Second, Lascaux is thereby misappropriated by the anthropological machine, used to draw yet another distinction between those who are and are not of our sort, to change complex and overlapping affinities and differences into a distinction whose central purpose has always been precisely to deny the possibility of any ethical responsibilities to those deemed other, to secure human dominion over the world.

Agamben (2004, 37) argues that every iteration of the anthropological machine operates by "means of an exclusion (which is also already a capturing) and an inclusion (which is always already an exclusion)," which together confirm prevailing presumptions about the features taken to distinguish the "fully" human. In the case of Bataille's analysis of Lascaux, the cave artists are *included* in the human order by *denying* that their ancestral precursors, and also their more distant relatives but relatively near contemporaries, the Neanderthals, were fully human. This exclusion of "rudimentary man" (Bataille 2005, 155) is, however, dependent on these same beings having neomorphic human features, *Homo faber*'s ability to think (149), the Neanderthal's ability to "react *humanly* before death" (150). They are thus deemed to occupy a zone of indeterminacy, neither animal nor human; they are set apart as "incompletely human." Bataille (2005, 75) also sets apart Lascaux's creators and contemporary "primitive" peoples from our kind on the basis that they (still) experience a love for animals, whereas "for us, animals are things." Humanity was, he claims, first born from the "poetic animality found in the caves" and then born again "from it by founding its superiority on the forgetting of this poetic animality and on a contempt for animals—deprived of the poetry of the wild, reduced to the level of things, enslaved, slaughtered, butchered" (76).

To the extent that he can be read as offering a teleological (temporal) compromise with the absolute (atemporal) distinctions, the either/or decisionistic logic, which the anthropological machine strives to install between human and animal, Bataille seems to remain wrapped within historicism's (and, in his particular case, Hegel's) shroud.[6] Such historicist teleology, the foretelling of a linear history, and the recounting of a now irretrievable past state of affairs, serves to justify and confer a degree of permanence on the anthropological machine's decisions regarding contemporary humanity's separate and dominant status. It provides a form of philosophical naturalization that gives the appearance of removing politics from the question of who now counts as human and what now counts as thing. Bataille's compromise takes this form: We can no longer see animals as kindred since we are no longer primitive—and yet the obvious question this compromise raises is how this assumed but unspecified "we" got from there (the poetic animality of Lascaux) to here (the contemptuous enslavement and butchering of the nonhuman world). (The complementary geographical question concerns Bataille's synchronic separation of contemporary primitive peoples—who in actuality have no relation of primogeniture to our (modern-Western) selves—from ourselves).[7]

In fact, any history detailing this emergence of supposedly essential human characters threatens the proliferation of zones of indeterminacy (of populations regarded as almost or not-quite-fully human beings) up to and including the present day.[8] This is clearly suggested by Bataille's sometimes injudicious applications of the term *primitive,* which simultaneously denotes a degree of relatedness and of separation. On the other hand, given the absolutism required by the anthropological machine, the decision to recognize the Lascaux people as nascent humanity can only work by recursively according them all of the qualities presumed necessary to be fully human from the beginning. Thus, in a way quite at odds with his accounts of "a consciousness less intricately confined by human pride . . . a humanity that did not clearly and distinctly distinguish itself from animality, a humanity that had not transcended animality" (55), Bataille also claims that "the die is cast from the outset" (79) we cannot doubt that Lascaux man [sic] already had the "sense of superiority and pride that distinguishes him in our day." This pattern of conflicting attributions reoccurs throughout his writings on prehistory.

The determinate decision required by the anthropological machine thus generates irreconcilable tensions within Bataille's accounts, especially around the status of the people at Lascaux. To the extent that he adopts a quasi-Hegelian teleology and is himself responsible for making a decision that the people at Lascaux comprise the first true humans, Bataille's work is compromised by these anthropological machinations. However, as Derrida (1995a) argues, Bataille's work is by no means a straightforward application of Hegelian principles but concerns itself precisely with that which escapes or cannot be entirely contained within the restricted economy of the Hegelian dialectic. He concerns himself with that excess which is never completely or entirely negated and conserved within each succeeding synthetic step, the remains deemed senseless, meaningless, and useless by the logic of human progress, a logic that they nonetheless continue to haunt.[9] As Kendall (2007, 96), editor and translator of Bataille's writings on prehistoric art, elsewhere notes, "A central fantasy of Western civilization has been that the entirety of the world and of human experience can be made useful and can be explained rationally. Bataille writes to reveal the madness of this fantasy." (This is also why Bataille's own work and life are deemed irrational and infectious by those shielded by this fantasy of a restricted economy, of a closed system of purportedly progressive exchanges.)

Perhaps, then, it might be argued that Bataille is not, despite appearances, simply conforming with the operation of the anthropological machine but also revealing, even reveling in, its residual contradictions, illustrating through his own contradictory claims about prehistory those zones of indeterminacy that are in no way resolved by announcing the sudden emergence of a truly human form. Certainly the tenor of his work as a whole makes this argument plausible, as does Bataille's ending his article on "the passage from animal to man" with a quasi-Hegelian speculation about what will happen to humanity and animality at the end rather than the beginning of history, when human subjectivity is no longer opposed to worldly objectivity. The contemplation of this synthetic, posthistorical disappearance of humanity, when humanity's "pretension" to set itself above a world of things will "cease being *clear* and *distinct*" (Bataille 2005, 80), will later provide the starting point for Agamben's (2004) reflections on man and animal and his critique of the anthropological machine.

However we choose to interpret Bataille, the startling originality of his analysis offers us other dialectical possibilities, not just by exploding the currently dominant ethicopolitical imagery, the restricted economy of an *intra*specific (human) social contract (see chapter 3), but in, however unintentionally, offering a critical perspective on the anthropological machine itself. If we read Bataille against Bataille, if we pay attention to the creative tensions in his work, to the moment of decision between proper and improper humanity that he can never actually make "clearly and distinctly," then different, ethicopolitical possibilities emerge as Lascaux's images collide with the now. We can recognize that the birth of humanity—at least insofar as it operates as a regulatory idea(l), as a category of beings separated from and dominating other beings—is not an irreversible past event described by the anthropological machine but a continually renewed and altering creation of its machinations: it is both metaphysical and politically partisan. As an initial thesis we might suggest then that the modern anthropological machine makes humanity recursively, in its own image, as an act of sovereign power, an act declaring human dominion over the world.

The full implications of this statement are not yet obvious; however, its acknowledgment offers the possibility of an ecological, ethical, and political critique of human sovereignty, of humanity's self-awarded distinction from and dominion over the natural world. Bataille's work suggests a congruency between the self-actualizing powers behind the decision that confers human status and confirms the domination of nonhuman nature. Only they who define themselves as fully human thereby grant themselves the power to decide who is excepted from this category, who is placed in a state of indeterminacy, who or what is to count as less than human. In Bataille's sense, the origins of humankind and of human predominance are not biological but dependent on an inherently ambiguous act of self-creation/self-distinction ex nihilo— from nothing and in the face of nothing, that is, in the face of death, human mortality, and today the realization of the inevitable eventual extinction of all human being(s). There is then also an elusive affinity, yet to be enunciated, between the people of Lascaux who, Bataille claims, "only wanted superiority over death" and those who today stake a claim on the basis of their humanity to preside over matters

of life and death for all other living beings. To the extent that modern Western society still seeks a clear distinction between human and
animal, a distinction that, by Bataille's interpretation, the people of
Lascaux did not clearly make, and that many contemporary societies
do not clearly make, this is not indicative of our greater humanity but
of being caught up within a socially particular system of thought dependent on the anthropological machine. This in turn suggests that the
operation of this machine is by no means necessary, all-encompassing,
or the final result of an inevitably progressing history.[10]

Ecological and Theological Dominion

In Benjamin's sense, the prehistoric images in the Lascaux caves can
conjoin "what has been with the now" (our current ecological situation), constituting new constellations of meanings and opening novel
political possibilities. (At least they have this potential so long as we resist temptations to employ these same images as vehicles to reinstall an
anthropological metaphysics of human origins and/or make them subject to forms of historicism.) Such ecopolitical possibilities emerge from
the ineradicable ambiguities and inescapable choices that are coeval
with feeling and recognizing affinities, differences, and ethical concerns
for a more-than-human world. Attending to the ethical quandaries expressed in the image of the bison's death subverts the metaphysics and
politics of the anthropological machine, challenges the modern constitution, and resists the claims of human sovereignty to be able to install an
unchallengeable hierarchic distinction between humanity and world.
An ecological politics takes sustenance from, and seeks to sustain, such
ethical concerns and involvements. In this sense, an ecological ethics is
the lifeblood of an ecological politics.

But even those sensitive to charges of historicism might still seek
to ask how we got from there (Lascaux) to here (our current ecological situation). Many environmentalists have certainly tried to trace
how more-than-human ethical involvements come to be dismissed,
ridiculed, and excluded (although they too have sometimes been captivated by the search for originary and decisive moments and/or been
tempted to reformulate history as a progressive worldly disenchant-
ment). The presumption of human dominion over nature, so integral
an aspect of every dominant modernist ideology, was, at least initially,
widely criticized as a key aspect of an ensuing ecological crisis that had

been long in the making. Although this critique, which had its heyday in the late 1960s and early 1970s, has largely been laid aside rather than addressed by contemporary environmental reformists, critics whose work has informed radical ecology have traced anthropological genealogies back through various discourses seeking to justify humanity's distinctive status: back to the dualism of Descartes and its separation of the thinking (human) mind *(res cogitans)* from mere matter *(res extensa)*, to the mechanistic (and masculine) experimental program of Bacon's *New Atlantis* (Merchant 1989), and even further back to ancient Greece and the West's Judeo-Christian heritage.

These accounts vary immensely in the subtlety of their analyses and the extent to which they recognize the interplay of ideas, social practices, politics, and ecology. They nevertheless bear witness to the continual production of varied iterations of the anthropological machine and associated discourses of human dominion obsessed with delineating a point or principle of original separation. Such original moments or principles are invoked to justify the ethical exclusion of the more-than-human but are mythic in Eliade's (1987, 95) sense that the purpose of myth is always to "proclaim what happened *ab origine*" as an "apodictic [sacred and hence incontrovertible] truth." The metaphysics and politics of human distinction and dominion are, from their very inception, inextricably caught up with the production of myths. It is no accident that those seeking "The Historic Roots of Our Ecologic Crisis," the title of Lynn White Jr.'s (1967) extraordinarily influential *Science* article, have often located them in the West's earliest myths.

Unfortunately, White's article was far from subtle in the ways it interpreted the admixture of metaphysics, ethics, and politics in biblical narratives and ancient Greek philosophy. He argued that the origins of our ecological crisis lay in the ideological sedimentation of a myth of human dominion based in the Judeo-Christian tradition of a God who created humanity in his own image to have "dominion over the fish of the sea, and over the fowl of the air . . . and over all the earth" (Genesis 1: 26). Humanity, though originally formed from clay, was thereby set apart from and above a natural world created only to "serve man's purposes" (White 1967, 1205)—a dominance established and exemplified by Adam, the first (hu)man, naming the animals. As "the most anthropocentric religion the world has seen" Western Christianity set the scene, says White, for the uninhibited exploitation of the natural world for human ends. Incidentally, the article also notes in passing the extinction, by 1627, of the European auroch, the beast whose images adorned Lascaux's walls.

Of course, from a Judeo-Christian perspective, humanity's ecological dominion is based in God's creative vision for the world rather than being something humans have simply claimed for themselves. The decision was God's, and yet Genesis also relates how the current human condition is consequent upon an initial human decision (and on subversive serpent wiles) to acquire self-knowledge, a consequence of which is to undertake a life in the constant face of death—the inevitability of human mortality. Here, contra White, we can recognize more subtle ways in which worldly concerns with life and death actually become mythically, metaphysically, and anthropologically invoked, for Genesis suggests that to be human, one of "our" kind, is from the beginning to take responsibility for our (knowledge of) death, although it also engenders, as at Lascaux, attempts to deflect this responsibility: the fall will also be used to expiate/justify the death of animal others in order to maintain human life (as Adams [1990, 112–13] notes, "It was commonly presumed that the Garden of Eden was vegetarian").[11] This mythic originating moment, as in Bataille's interpretation of Lascaux, also marks the release and recognition of desire and shame, of knowledge of good and evil (ethics), and also the beginning of human labor that would, initially out of necessity, transform the world to suit human intentions.[12]

That the anthropological machine is already at work here is evidenced by woman, as is so often the case, being placed in a zone of indeterminacy, simultaneously included and excluded. She is created together with man in the first Genesis narrative where "God commands man-woman to command the animals" (Derrida 2008, 16), but in the differing second narrative, where she is created *after* man, God "lets man, man alone, Ish without Ishah, the woman, freely call out the [animals'] names." Created alongside man in God's image, she is perfectly human, and yet her role is to exemplify human imperfection. And since this story is couched in terms of an all-too-human fall from innocence and grace, rather than a Hegelian teleology of spiritual ascendance, she also appears after, rather than before, the first perfect, otherwise incorruptible, still innocent man. She is the creation of the first excision, the spare rib, the subjective object of desire, the corrupting afterthought, she who engages in creaturely conversations, she who on their mutual expulsion from the garden will be declared subservient to the dominion of this first man.[13] And while the anthropological machine cannot work without (placing) her (in this indeterminate state)—she is, after all, Eve, the mother of all who live (Genesis 3: 20)—she is also the first to rebel against the claims of sovereignty. And she, like Adam, already knew, even before tasting, that one of the seeds of self-knowledge was the recognition of life and death and the ethical quandaries that can arise only through such knowledge.

Perhaps in his eagerness to identify the origins of human dominion, White misses how, even in such an anthropocentric myth, ethics, ecology, and individual actions still appear as disobedience to the principles dictated by sovereign power. An ethical/ecological ambiguity exists even here, at the mythic beginning of history, since it is only due to this disobedience (which leads from innocence to ethics and ignorance of death to knowledge of mortality) that people find themselves exiled by sovereign power (thrown into the world). But also, only thus do people *find themselves,* as beings intimately and *ethically* concerned with this living (and far from mythic) natural world.[14]

Unfortunately, the debate sparked by White's article never engaged with such subtleties or with the principle of sovereignty itself, still less with its contemporary political manifestations. Rather, the presence/absence of (ecological) ethics became glossed in terms of a debate between visions of Judeo-Christianity as dominion theology and those who championed a more pastoral model of metaphysical/political power. Attfield (1991, 27), for example, argued that the "biblical dominion of man is no despotism" but should be interpreted in conformity with the Hebrew concept of a monarch, as someone in turn answerable to God. This accords with Glacken's (1967, 166) much earlier exegesis, whereby humanity is granted only a derivative power—a form of earthly stewardship. The image associated with this notion is that of the "pastorate," the good shepherd and his flock. However, as Passmore, another early contributor to these debates argued, stewardship too can be understood, as it is by Calvin, as "the rule of the elect over the reprobate" (Passmore 1974, 29) and as indicating a master–servant relationship. And as critic Neil Ascherson (2006, 11) recently remarked, the "notion of human trusteeship for the natural creation . . . is benign in intention and often in consequences, and yet perpetuates the claim to a sovereign, supra-natural status for the human race."

However well intentioned it may be, stewardship's critique of ecological despotism is not a critique of dominion or sovereignty nor of the unique position allotted to humanity. While it seeks to broaden human responsibilities in terms of recognition of a sacred trust, its strategy is to argue the need for human compliance under a superior, overarching sovereign power, albeit one displaced into the supernatural realm. Such "service is the will of him who charged us with dominion; its purpose is to preserve, enhance, and glorify the creation, and in so doing, to glorify the Creator. In short we are *stewards* of God, managers of this particular part of his household" (Wilkinson 1991, 308). This vision of humanity as household servant/manager (and those promoting stewardship constantly recall that the origin of the term *ecology* itself lies in the ancient Greek *oikos*—household) seems strangely appealing to many. (And the model of the pastorate can also be understood, as Foucault [1994] points out, as a precursor to modern forms of biopolitical governance, the management of populations [flocks] in matters of their life and death.) Stewardship remains a fundamentally theocratic and paternalistic model wherein responsibilities for nature are actually inseparable from subservience to God and potentially, depending on how directly or indirectly the relation to God is theologically envisaged, to God's (self-proclaimed) representatives on earth.

This is not to say that such debates are without practical ecological or ethical consequences, especially given the prevalence of current forms of dominion theology associated with the rise of fundamentalist Christianity and far right politics in North America. Basing their self-righteous certainties in the same biblical quotations identified by White, these fundamentalists regard environmental destruction as a symptom and symbol of the end times and even, in some cases, as something to be welcomed as prefiguring the rapturous exit of the saved from all worldly involvement (Hendricks 2005). To this extent, godless communism has been replaced by a this-worldly environmentalism now regarded as no less threatening to the (holy) spirit of American capitalism. Any evolutionary connections between humans and the natural world are vehemently denied. But while the idea of stewardship might counter at least some of these extreme claims, the problem is that questions of ethics, politics, and ecology are still subsumed under differing theological interpretations of myths as sacred truths and metaphysical certainties.

In terms of an ecological ethics and politics, such myths need "secularizing" in Vattimo's (2004, 57) sense of the weakening *(Verwindung)* of their claims to absolute truth or validity (see also endnote 16). For

Vattimo (a Christian who has been a key contributor to recent debates concerning the death of God, see Caputo and Vattimo 2007; Derrida and Vattimo 1998), this secularization is itself an ongoing tendency in Western thinking that he explicitly associates with the "dissolution of the divine right of all forms of authority" (2004, 32). He argues that a reflexive understanding of the implications of "the death of God" entails the realization that all now face having to take ultimate responsibility for their ethicopolitical actions, values, and associations—since these can no longer be grounded in a naïve assumption of a shared cultural understanding of divine authority.[15] This secular tendency is not restricted to criticizing the metaphysics underlying religious authority but would also incite a call to further secularize, that is, critically weaken, religion's secular equivalents—for example, the attempt to replace God's overarching presence with historicist concepts like the inevitability of progress.

Of course, Vattimo's secularization thesis can itself be interpreted as yet another form of quasi-theological and historicist grand narrative "selling another metaphysical bill of goods this time under the name of demythologization" (Caputo in Caputo and Vattimo 2007, 82). But if we take secularization as simply a hermeneutic possibility rather than a more or less inevitable historical process, a possibility accompanied for Vattimo by the (kenotic) translation of ideas of God into a this-worldly (rather than supernatural) setting, then this might provide a demythologized sociotheoretic understanding of the ethicopolitical implications of Eve's original decision, where we recognize that such ethical quandaries are, and always have been, incumbent upon an ability to take ethical responsibility for our actions. Ethics is choosing our own difficult paths in the face of sovereign power's dictates. This secularizing incarnation, this reflexive internalization and recognition of what might be termed a divining (hermeneutic and human potential) rather than a superior divine power is potentially immensely liberating and yet, precisely because of this, also leaves people abandoned to constitute ethics and politics together in this far from perfect world. In this sense, then, if we are to be stewards of nature, this would have to be because, through ethical and political deliberation, we have ourselves come to this conclusion, not because we have prescribed duties owing to a supernatural being.

Just occasionally those involved in debates between ecological dominion and stewardship recognized that the texts being interpreted

were actually indicative of the incomplete triumph of the anthropological machine, its inability to entirely eradicate the ethical moment
opened by and through the realization of the worldly implications of
being human. Thus Black (1970, 40) argued that although "man has
seen himself as licensed to dominate the earth, he has not been able to
accept completely and wholeheartedly this place in the hierarchical arrangement God : Man : Nature." For Black, the declaration of human
dominion sat uneasily with many experiences and understandings of
nature and, in particular, of animals' lives. His concern was precisely
to elucidate the ambiguities and oppositions he considered inherent in
biblical creation legends rather than argue a definitive interpretation.[16]
Such ambiguities, he suggested, reveal how ancient and continual this
troubled relationship with animals is, how ethics has, from the beginning, constantly subverted the kind of claim to dominion that regards
animals as no more than freely utilizable things.

Bataille (2005, 78) drew a similar inference from the image at
Lascaux: "The constant ambiguity of humanity is originally linked to
this duplicity with regard to animals," that is, their treatment as both
real, kindred beings and as mere things. The "point of the evocation
[the drawing of the animal in the cave] is overtly the reduction of the
real being to a possessed thing. In fact, the evocation is also, in an essential way, an excuse: in addressing the animal, the human predator
asks forgiveness for treating the animal as a thing, so that he will be
able to accomplish without any remorse what he has already apolo-
gized for doing."

Interestingly, Passmore too suggests, albeit only in a footnote, that
claims of humanity's dominion over nature were employed to legitimate, rather than instigate, practices of "mastering" and "subduing"
nature that were actually already well underway by the time Genesis
was compiled. Genesis, says Passmore (1974, 7), merely "salved his
[humanity's] conscience," which, if true, also suggests that ethical concerns about nonhuman nature underlie even this most anthropocentric
image of human origins no less than they do those of Lascaux.

Plato's Cave and the Polis: Myth and Ethos

Even if these myths recounting human origins and justifying ecological dominion are understood as an early iteration, indeed on White's
reading, the archetype, of what we have chosen to call the anthropo-

logical machine, they are still shot through with ethical ambiguities that can weaken (in Vattimo's sense) their theocratic and metaphysical pretensions.

But what of philosophy and especially the ancient Greek philosophies that White and other environmentalists also identify as ideological precursors of our contemporary ecological crisis? For example, Passmore (1974, 189) too claimed that the idea of "man as despot" did not originate in Genesis per se but only in its conjunction with certain strands of Greek philosophy, especially those that followed Plato in condemning the "sensuous" life of humans, a trait he considers archetypically authoritarian. Of course, *sensuous* here, as Passmore makes clear, has few permissive or transgressive connotations (as it had for Bataille), but it does at least recognize the dangers in defining human perfection, the properly human, in disembodied, otherworldly, and overly rationalistic terms. The "attempt to be 'super-human' by rising totally above sensuousness issues . . . is a way of life, no less impoverished, no less 'sub-human' [than the purely sensuous life] and is utterly destructive, into the bargain, of man–nature relationships" (Passmore 1974, 189). "Only if men [sic] can first learn to look sensuously at the world will they learn to care for it. Not only to look at it, but to touch it, smell it" (189) and, as Eve might add, "taste it."

Passmore certainly identifies the poles of a recurring debate here, but there is a sense in which, even as he writes, his ethical and ecological critique relies on yet another version of the properly, fully, or more perfectly human: those who are neither "super-human" rationalists seeking to separate themselves from and dominate nature, nor "sub-human" sensualists losing themselves in an Edenic "primitivist view" of a nature already "perfect as it is" (1974, 38). This "half-way house" Passmore hopes to inhabit could easily be used to install the anthropological machine in yet another guise. So any radical ecology needs to bear in mind that *to interrupt the anthropological machine, an ethical concern for nature would itself have to be an expression of a relation not predicated on whether or not such a concern is properly human.*

That said, the key ecological point for Passmore, as for White, is the way that Plato's philosophy encapsulates a metaphysical idea(l) of a humanity dialectically progressing toward epistemological, moral, and political perfection through a process of rationally separating themselves from, yet placing themselves in a position of sovereign power over, the material (but illusory) world. To this end, Plato too, as might be expected, employs not only philosophy but mythology. The two are entwined so closely that they are often inseparable, and yet, precisely because of Plato's subsequent influence, it remains vitally important to trace at least some of the ways in which ethics, politics, and ecology might exceed and subvert this mythic/metaphysical framework that seeks to bind them.

Ironically, given Bataille's interpretation of Lascaux, Plato's mythic/ philosophical cave represents the realm of a still unenlightened humanity caught up in worldly appearances, mistaken in their felt affinities with the shadowy, flickering forms of nature that are represented by firelight on the cave's wall. The cave is no longer a mysterious source of intimate involvement in the (plural and manifold) dialectics of existence, the movements to and fro between human and animal life and death. Instead, it is envisaged as a material prison in which (some) people seem trapped, drawn to and confounded by images that bear little relation to Plato's newly defined metaphysical "reality." The only animals here are shadows of human-produced "shapes . . . wrought in stone or wood and every material" carried, along with "implements of all kinds" (Plato 1963, 747 [*Republic* VII, 514c–15a]) behind the backs of the fettered populace, their associations with a living presence reduced to puppet theater.

Plato's philosopher, as the embodiment of reason and by dint of the *force* of reason, drags those who can be freed from worldly illusion into the light of a truth newly discovered by this same (single and universal) dialectic, that is, a movement to and fro in thought, between discussants in dialogue, but also within the philosopher's own mind. Thus, even as dialectics are defined, their very nature is shifted—no longer the complex material, mimetic and metonymic, intermingling of differences and affinities between nature and culture but an abstract (nonsensuous) process (that too might initially appear as a restricted economy) mediated within human language and dependent on its presumed ability to rediscover the rational forms, the immutable essences, that underlie language and appearance. In a letter to the supporters of the assassinated Dion, tyrant of Syracuse, Plato (1963, 1591 [Letters VII, 344b], my emphasis)[17] describes this process: after "practicing detailed comparisons of names and definitions and visual and other sense-perceptions, after scrutinizing them in benevolent disputation by the use of question and answer without jealousy, at last in a flash understanding of each blazes up, and the mind, as it exerts all its powers to the limits of *human* capacity, is flooded with light."[18]

This light, though, is only metaphorically that of sunlight, just as the "affinity with justice and all the other noble ideals" that Plato (1591 [344a]) denotes as properly human seem estranged from ethical affinities/ambiguities of the kind expressed by Lascaux's art. This is because ethics too is now reenvisaged in terms of language and essences, of moral concepts, the knowledge of which remains elusive if the person concerned is (as most are, according to Plato) "naturally defective," lacking a "natural intelligence"—defined precisely in terms of their inability to employ reason's dialectical process. The consequence is that most people, lacking either a natural affinity with and/or an intellectual ability to grasp the metaphysical *form* of justice, "will never any of them attain to an understanding of the most complete truth in regard to moral concepts" (1591 [344a]). Since they fall short of the supposedly natural "limits of human capacity," they are exempted from being fully human (just as Plato thinks they should be exempted from any part in ruling the polis). The "stupid and unretentive" (1591 [344a]) join barbarians, slaves, those made of "base metal," and women, in zones of indeterminacy, placed there by the decisive power and self-justifying logic of philosophy, a new sovereign power that, Plato suggests, might be reflected politically in the ideal form of the *Republic*'s "philosopher-kings."

This anthropological machine operates throughout Plato's works. For example, in the *Theaetetus*, he has Socrates define philosophy precisely in terms of "*what man is*, and what is *proper* for man's nature to do and suffer, as *distinct* from the nature of other *things*" (Plato in Cavarero 1995, 53, my emphasis). And as Cavarero points out, this definition occurs in the context of the story of the Thracian maid, who as both woman and household servant is doubly relegated to a zone of indeterminacy but still has the effrontery to laugh when the ideal man, the philosopher Thales, caught up in contemplating the heavens, fell instead into the depths of a well. "There is indeed a very good reason for the maidservant to laugh" at the philosopher's sudden and precipitous loss of dignity, says Cavarero (1995, 53), "for Socrates the Thracian servant, like any other woman, has her real, true being (though disempowered and inferior) in the idea of man" generated and supposedly epitomized by such philosophers. According to this dominant philosophical perspective, "Women do not constitute the 'other sex' of the human species, but rather a sub-species." The maid's laughter (so disruptive to absolute authority's claims) expresses dramatic irony concerning the true philosopher's careless failure to recognize the worldly realities he has excluded by and from his metaphysical ruminations.[19]

Paradoxically then, the political sovereignty accorded to Plato's philosopher kings depends on decisions made about who is properly or most fully human by those who have already *defined themselves* as such. This decision places obvious constraints on politics as such and also delineates the bounds of a moral community where those inhabiting the city's various zones of indeterminacy are, because of their "incompletely human" status, deemed incapable of making their own ethical decisions. Rather, their ethical responsibility is redefined in terms of moral obedience to sovereign power—they must follow the moral regulations laid out for them through the philosopher-kings' role as lawgivers. As Passmore (1970, 44) notes, "Most men cannot achieve what Plato significantly calls 'philosophical' goodness, for him the only true goodness. The most they can hope for is civic goodness, the sort of goodness they can acquire by obedience to the rules laid down for them by the philosopher-kings."

The classical description of Plato's *Politeia,* then, presents an ideal state ruled by ideal humans. It provides a metaphysical myth of philosophical dominion over politics and the rational ordering of an ethos. As Gadamer (2000, 48) argues, the *Republic* is the earliest comprehensive example of a concern with "the problem of good and its concretization in an ideal city. Yet one must recognise that the ethos of Plato's *Politeia* has a utopian dimension. . . . This ethos appears . . . in such a way that everything there is regulated. There it is nearly impossible to do anything that is evil or abnormal." This seems like an ethical as well as political despotism, one which dictates behavioral norms and enforces a common *doxa* requiring absolute obedience to the laws and restricting the freedom to institute, question, or on occasion alter those laws to the legislators themselves.

Interestingly, as Passmore also notes, Plato often portrays the polis in terms of stewardship, employing pastoral myths and dialogues as in the *Republic* (see Plato 1963, 593 [1: 343]) when Socrates invokes the image of the care of a shepherd for his flock as a metaphor for good political rule. The imperfect human flock are deemed to be dependent for their well-being on those philosopher-kings who have awarded themselves the status of household managers/shepherds on the basis of their control of the dialectic and the knowledge (including the ethical knowledge) its possession guarantees.

And so this situation raises again questions about the differences, if there are any, between dominion and stewardship—albeit this time as more of a philosophical issue concerning human politics than a theological issue concerning dominion over the natural world. That said, this relationship, and the link between dominion over nature and the polis, is nowhere made clearer than in Plato's *Statesman* where, interestingly but in no sense accidentally, the art of the ruler is dialectically defined by the Stranger, with the assistance of a young Socrates, in terms of a series of classificatory decisions (the philosophical equivalent of Adam's naming the animals) concerning the natural world, beginning with that between living animals and nonliving things, then tame and wild animals, then tame land and water animals, then flying and walking land animals, then horned and hornless walking animals, then gregarious (herd) and nongregarious animals, and finally four- and two-footed herd animals. "And so," says the Stranger, "we reach the object of our search, namely, statesmanship or kingship, which is another name for statesmanship" (Plato 1963, 1032 [267c]). The role of the (philosopher) king is as the herdsman of living, tame, featherless, hornless, gregarious bipeds, that is, of humans (now taxonomically distinct from, though still placed in relation to, all other beings).[20] This passage perfectly illustrates the way in which this narrow (nonsensuous, abstractly rational) dialectic is deployed in the interests of those most adept at using it to define their political power while simultaneously instantiating hierarchic anthropological distinctions between humans and the natural world.

Again it is tempting to claim that Plato's *Polis* might simply exemplify a form of despotism masquerading as an ethically concerned stewardship of the city's populace for their own well-being. In the *Republic,* Thrasymachus, as Passmore (1974, 9) notes, had raised precisely this "critical ambiguity" in Socrates's pastoral imagery, asking him whether he thinks that shepherds fatten and tend their sheep for the benefit of the sheep themselves or the shepherd's own profit. Yet one need not be a moral skeptic like Thrasymachus to see the potential problems in an image of stewardship that depends entirely on an analogy between the ruled populace and herd animals. The Stranger argues that statesmanship is not a matter of nurture but of concern (Plato 1963, 1042 [276d]) of the "'responsible charge' of a

whole community" (1042 [276b])—"what other art can claim to be the
art of bearing sovereign rule, the art which bears sovereign rule over
all men" (1042 [276b–c]). It is also, he claims, not a matter of tyranny
but of free acceptance. "Tendance of human herds by violent control
is the tyrant's art; tendance freely accepted by herds of free bipeds we
call statesmanship" (1042 [276e]).

Of course, the extent to which Plato's theories allow for free acceptance is questionable, while the fact that his Stranger classifies humans as herd animals is not. Plato's version of the "pastoral modality
of power" (Foucault 1994, 300) once again illustrates that the difference
between a dominion and stewardship is by no means as politically substantial as stewardship's proponents might suggest. While the despot
bans politics as being contrary to their own absolute claim to sovereign power, the pastorate favors the biopolitical reduction of politics to
the model of household management. Rather than a democratic and
dissenting polis, we have the paternalistic control and management
of populations on the basis of their predefined needs. Stewardship, of
course, does this on the basis of its claim to be operating under the
auspices of a higher ethical power that, properly understood, guides
the rulers' concerns for the well-being of those ruled.

On the theocratic model, this stewardship is held under the auspices
of god, but Plato is already on the way to secularizing such claims. To
be sure, on occasion these higher powers are referred to by Plato in
mythical terms as gods, but these gods seem to be an ambiguous and
additional, rather than an integral, element of Plato's political dialectics. In the *Statesman*, Plato (1035–36 [270c–d]) explains this strange
presence/absence of a supernatural ruling power in terms of yet another mythic narrative. The Stranger relates to Socrates an account of
how, at a moment of "cosmic crisis" when "there is widespread destruction of living creatures other than men and . . . only a remnant of the
human race survives," the gods turn time back on itself to the very
beginning of human existence. The orderliness of nature is thereby renewed, the old become young, the dead are returned to life from the
earth where they were buried, and each generation successively fades
into nonexistence until we come to a holding moment when, once again
released by the gods, time will begin to flow forward again.

This holding moment is a suspended moment of transition when,
paradoxically, our "earliest forbears were the children of earthborn parents" (1036 [271a]), that is, of the last resurrected dead, newly reborn

from their earthly tombs. These people (like those of Bataille's Lascaux) are envisaged as our earliest but still indeterminate forbears, situated both at the beginning of the "present" rotation of time and giving birth to our kind, to mortal men, but also themselves born from the very last of the earthborn parents that preceded, and recede from, them. Since they are also the last generation of the resurrected dead, then these same forbears are also, though in a different way, born from themselves, self-produced and self-defining, now in the face of their very own deaths.

Before this moment, under the rule of Cronus, the earth experienced a primitive Golden Age, "when all good things come without man's labour," when "over every herd of living creatures throughout all their tribes was set a heavenly daemon to be its shepherd . . . providing for the needs of all his charges. So it befell that savagery was nowhere to be found nor preying of creature upon creature, nor did war rage" (1037 [271d–e]). Here people were able to "converse with the animals as well as one another" (1037 [272b]); "they had fruits without stint from the trees" (1037 [272a]) and "disported themselves in the open needing neither clothing nor couch" (1037 [272a]). But this familiar pattern changes when, released once more into the forward flow of time and abandoned by the "formative action of external agents" (1039 [274a]), the universe is left to "take sole responsibility and control of its course" (1039 [274a]). In this situation, every kind must be responsible for itself and "bereft of the guardian care of the daemon who had governed and reared us up" (1039 [279b]), humans, having also lost the ability to converse with animals, are left at the mercy of savage beasts, forced to defend themselves and fend for themselves through the Promethean gift of fire and technology. Once again, here at the origin of humanity, we have the same anthropological distinctions emerging but also the same pattern of humanity being understood as having to take responsibility for its actions (together with the same shifting of guilt and responsibility in relation to nonhuman others). And the only justification that the Stranger gives for why this world should be preferred to the primitive Golden Age where "the happiness of the men of that era was a thousandfold greater than ours" (1038 [272d]) is that this fails to satisfy the insatiable human desire for knowledge.

Politics, then, for Plato has to be understood in this context. It is an invention made necessary by humanity inhabiting this imperfect world. It is an attempt to take responsible charge of a community abandoned by the gods to its own largely destructive devices. In the Golden

Age, the "deity being their shepherd mankind needed no political constitution" (Plato in Foucault 1994, 306). This, says the Stranger, shows
that the pastoral metaphor rightly applies only to the rule of a god, not
a mortal, to a realm under a perfect order. In an age that is "*said* to
be under the government of Zeus" (1037 [272b], my emphasis) rather
than Cronus, Plato's philosopher-kings are abandoned—left to decide
for themselves how to weave the political fabric of the polis together,
helped only by such illustrative myths and the dialectic.

In his analysis of this text, Foucault (1994, 307) suggests that Plato
impugns the very idea of the pastorate as a model for politics (see also
Kalyvas 2005).[21] But this is not exactly so, since the Stranger explicitly
states that "what was said [about the role of the king as shepherd] was
true, but it cannot be regarded as the whole truth" (1040 [275a]). In fact,
there is a way in which Plato is certainly happy to retain this pastoral
metaphor, using it in other texts (see later in this chapter) and giving
it voice through the mature Socrates's own words (see earlier in this
chapter). However, what Plato has effectively argued here is that the
very idea of stewardship under a god will work only for a world where
gods are present, where god is not yet dead and decisions are not actually made by human beings.

When it comes to human rulers, in place of a stewardship that would
be, at best, only a form of puppetry on behalf of a supernaturally benign
despot (a god), at worst a lie to mask the despot's all-too-human face,
Plato suggests an explicitly human form of stewardship empowered
by the dialogic achievement of an understanding of ethics (the Good).
For Plato, the philosopher-kings are themselves presumed to be acting under what Iris Murdoch (1970) calls *The Sovereignty of Good*.
The dialectical movement of the philosopher-kings toward this perfect
combination of Truth (knowledge) and the Good (ethics) is, Plato argues, the best way to ensure an approach toward human personal and
political perfection in accordance with the limits of human capacities,
that is, toward utopian ideals that always lie ultimately out of reach in
an imperfect world.

In effect, though, despite its ingenuity, this metaphysical solution
risks placing very real constraints on the way that ethics and politics
are both envisaged and exercised, constraints that are directly and indirectly justified through recourse to idea(l)s of the fully, or most perfectly, human, that is, through the operations of the incipient anthropological machine. The dialectic, having been separated and purified

by abstracting it from sensuous involvement in the world and making it the very mark of the properly (fully) human, is then reified as the principle of sovereign power wielded by the philosopher-kings. It also makes politics the means to an end rather than, as Arendt (2005) and Agamben (2000)[22] argue it should be, a means in and of itself (in Agamben's terms, a "pure means"), a realm of diversity and freedom of expression that is a constitutive feature of any actual community. What is more, depending on how the ideal of the Good is interpreted, Plato's solution risks, however unintentionally, making politics subject to an "ideocracy" wherein one truth supplants "the many relative truths that Socrates relentlessly sought to bring to birth by questioning his fellow citizens" (Kohn in Arendt 2005, xxvi). As Arendt (3) argues, insofar *"as action is dependent upon the plurality of men [sic] the first catastrophe of Western philosophy, which in its last thinkers ultimately wants to take control of action, is the requirement of a unity that on principle proves impossible except under tyranny."*

2 THE SOVEREIGNTY OF GOOD

THE MOTIVE FOR ADDRESSING PLATO'S WORK is not simply because of his subsequent philosophical and ideological influence. Even Western philosophy is far from being, as Whitehead (1978, 39) famously suggested, just "a series of footnotes to Plato." Still less is it to paint him as ultimately responsible for our current ecological crisis. It is not even to argue that metaphysics and myth should (or could) be entirely abandoned because of the political dangers they pose in conjunction with sovereign power. Rather, tracing the ways in which ethics, politics, and ecology are transformed, defined, and made subservient to such an overarching ("totalizing," and on some readings [e.g., Popper 1969] "totalitarian") philosophical system can also reveal ways in which they remain and reemerge in its very midst as subversive possibilities.

The vicissitudes of temporal existence, the unpredictability of politics, and the ambiguities of ethics lead Plato to formulate a heady mix of myth and metaphysics to justify the sovereign rule of those who, like him, have defined themselves (on the basis of their representing the timeless essence of philosophy and statesmanship) as properly human. Their sovereignty is exemplified in the (self-serving) decision to exclude those deemed improperly human from even practicing either philosophy or politics! This decision is also justified in terms of a biopolitical model of the pastorate—the stewardship and shepherding of state and people for their own protection and welfare. The real world is thereby, at least theoretically, made subservient to a metaphysical realm where everything is ordered according to rational (nonsensuous), overarching, timeless principles: life is reduced to something to be managed and preserved, politics to a matter of and for statesmanship, and perhaps most ironically, ethics is transformed from the irruption of worldly concerns for Others into an otherworldly sovereign principle. The stewardship ideal helps Plato retain an aura of ethical concern even as he seeks to install the Good as the original principle *(archē)* of political authority and as a sovereign limit on the exercise of politics.

Of course, we have still to consider the ecological implications[1] of the ways Plato's subtle admixture of myths and taxonomic exercises connect nature to politics and ethics even as they instantiate an early iteration of the anthropological machine. Indeed, political and philosophical authority is deemed necessary partly to ameliorate the chaotic consequences of our imbricate setting within the flesh of the world, to impose order and regularity through the restricted economy of the dialectic. A rule-directed life, conducted within the bounds of a city-state is, Plato believes, preferable to an undistinguished and indolent life immersed in nature, however pleasurable it might be. Certainly he thinks the philosophically administered life in the polis preferable to the anarchy he associates with the "bodily element" (Plato 1963, 1038 [*Statesman*, 273b]), that materiality which has been part of the universe's constitution from primeval times and which he blames for the world's decay, its mortality, its susceptibility to time's forward flow.

Blaming the materiality of the world for its own demise seems somewhat ironic, ecologically speaking, especially given that, at least occasionally, Plato exhibits awareness of the potentially destructive impacts of certain human interventions in nature. For example, in the *Critias*, Plato offers what might be taken to be another account of the Golden Age, although one now presented as early history more than timeless myth. Once again the gods are described as shepherds tending their human flocks, using persuasion rather than force to guide them, "so steering the whole mortal fabric" (1215 [*Critias*, 109c]) in a way more suitable to human intelligence. These gods, we are told, "produced from the soil a race of good men and taught them the order of their polity" (1215 [109d]). But the successors of these ancestral Greeks bring upon themselves what has been regarded as one of the first descriptions of an ecological disaster (see, e.g., Glacken 1967, 121; Coates 1998, 28). At first the soil of Attica "far surpassed all others" (Plato 1963, 1216 [*Critias* 110d]), so much so that "the remainder now left of it is a match for any soil in the world" (1216 [110e]). But this soil washed away so that "what is left now is, so to say, the skeleton of a body wasted by disease; the rich, soft soil has been carried off and only the bare framework of the district left" (1216 [111b]).

The original soil, says Plato, once ensured the percolation of water necessary to support many springs and provided for the growth of abundant forests, which in turn provided fodder for beasts and rafters to support vast buildings. But now, he notes, only a few trees and the spring's sanctuaries still survive. Plato blames these evil tendencies on the "constant crossing [of humanity's divine strain] with much mortality" and on the "human temper to predominate" (1224 [121a–b]), together with human thoughtlessness and forgetfulness.

While these surviving fragments of the *Critias* appear to present almost an ecological version of the fall, these concerns remain marginal in terms of Plato's conceptions of the Good and the Republic. Plato is no more the forerunner of an ecological consciousness than he is culpable for our ecological crisis. So any connections made to ecological politics must draw primarily on the more general ethical and political possibilities that might still be present within his work. We need to approach the questions of ecological ethics and politics and ecological sovereignty obliquely via the question of the sovereignty of the Good and via differing readings of Plato's works—readings that emphasize its critical possibilities rather than its (antipolitical) intent.

While it is easy to critique the *Republic* as if it merely sets out an antipolitical blueprint for a future society, Gadamer (2001, 84) argues that Plato's *Politeia* is actually utopian in the sense that it offers only "suggestiveness from afar." Plato never wanted or expected it to be realized as such because "the primary function of a utopia" is, says Gadamer, the "critique of the present, not the construction of whatever project being described in the work." To "suppose that Plato ever thought that the *Republic* was attainable would be to suppose him capable not merely of optimism or idealism but of sheer political *naïveté*" (Saunders 1984, 27–28). Or to quote Iris Murdoch (1970, 94), "Plato, who understood this situation better than most of the metaphysical philosophers, referred to many of his theories as 'myths,' and tells us that the *Republic* is to be thought of as an allegory of the soul. 'Perhaps it is a pattern laid up in heaven, where he who wishes can see it and become its citizen. But it doesn't matter whether it exists or ever will exist; it is the only city in whose politics [the good man] can take part.'"[2] Plato's *Republic* is, in other words, an ideal city and what this idea(l) sets forward is the way that politics might be guided by ethics, which is itself to be understood as a utopian ideal—the Good.

As a reading of Plato's intent, this understanding of utopianism is contentious—after all, except for the primacy given to philosophers, many aspects of the *Republic* seem close to the actual ordering of the Spartan state, and as Ernst Bloch (who certainly recognized the *Republic*'s utopian claims) argues, Plato's later *Laws* offer only a "burnt-fingered social utopia" (1997, 486) approximating a "police state" where even this difference evaporates. Nonetheless, a key question for radical ecologists, who also desire an ecological politics informed, and at least to some extent guided, by ethical concerns for the more-than-human world might be how to retain certain critical and utopian aspects of Plato's philosophy while resisting the antipolitical installation of ethics as a sovereign principle. After all, as Bloch (1995, 7) reminds us, "All freedom movements are guided by utopian aspirations." Might there be other metaphysically weakened ways of articulating ethics, ecology, and politics that, while retaining something of the utopian aspirations of this idea(l) of the Good, do not reduce its role to that of an ideocratically employed despotism—that is, as an overarching truth with which those subject to sovereign powers are expected to comply?

The following three sections begin to address this question, first by confronting questions concerning the metaphysics of ethical sovereignty, then by exploring possible links between the metaphysically weakened understanding of ethics that results from this confrontation with ecology, and finally by beginning to suggest an alternative model of a mutually informative relation between ethics and politics.

The Good: Plato, Iris Murdoch, and Emmanuel Levinas

The philosophy of Iris Murdoch explicitly addresses Plato's work in a way that has direct relevance to an ecological ethics. In *The Sovereignty of Good* (1970) and throughout her philosophical writings and novels, Murdoch engages in a form of metaphysical theorizing that often seems unappealingly anachronistic from modern perspectives. Her work invokes a quasi-Platonic conception of the relationship between forms (ideas) of the Good (ethics) and the True (how the world really is). Here the Good "refers us to a perfection which is perhaps never exemplified in the world" (93), yet, as in Plato's allegory of the cave, once we stand "in its light we see things in their true relationships" (92). Such a universal coincidence of the Good and the True is difficult to accept in a contemporary culture so dependent on the separation of (though constantly redefined distinction between) spheres of values and facts.

However, Murdoch's philosophy is, in many ways, an attempt to account for and dissolve this fact–value distinction without falling into a reductive ethical naturalism. And, while Murdoch is quite explicit about her immense debts to Plato and her metaphysical commitments in terms of her concerns with the nature of reality and ethics, her interpretation of Plato is itself radically "non-metaphysical" (93), if by metaphysical is meant (as many of her Oxford contemporaries intended) concerned with abstruse otherworldly speculation.

This, of course, is precisely how Platonic forms are often interpreted, although a different genealogy could be traced, running from the neo-Kantianism of Paul Natorp to the hermeneutics of Gadamer and the phenomenology of Emmanuel Levinas, that would emphasize, albeit in radically different ways, what Sullivan (1985, xii) refers to as the "hypothetical" rather than "objective" understanding of Platonic ideas. In this sense, Murdoch might be thought of as presenting an understanding of Plato that is more than usually secular in Vattimo's sense. On Murdoch's reading, Plato's texts provide mythic metaphors (like that of the cave) suitable for the philosophical task of understanding ethics (the Good), and his metaphysics is the means by which these metaphors are conveyed explicitly and systematically. Murdoch, again unlike many of her contemporaries, believed that philosophy needed such metaphors and such conceptual schemes—metaphysics in this more *mundane* sense—in order to ask important ontological and ethical questions about what the world is like. She thus tries to provide a reading of Plato that is neither dependent on belief in supernatural entities nor yet reducible to the kind of nonevaluative naturalism that claims to confine itself to empirical descriptions. The idea (form) of the Good is not, on Murdoch's reading, something we need to think of as opposed to or underlying *material* reality; it is not some extraworldly quality floating in another metaphysical dimension accessible only to philosophical thought. Rather, the Good is the idea—the understanding of the form taken by—ethics as such.[3]

Just what this means might be approached by explaining Murdoch's indebtedness to, and differences from, G. E. Moore (1922, 118), whose answer to this same problem—"What is good in itself?"—was unfortunately, in her view, to prove so damagingly influential to the subsequent course of ethical theory. Moore famously argued that the question of what good is can never be delimited by any description of any particular states of affairs, even where those states of affairs concern ethical appraisals. This is because, in Hepburn's (1995, 606) words, the "question [Is *that* good?] always remains open, and never becomes trivial. 'Good' resists definition or analysis; and the attempt to pin it down to an invariable specific content is, in Moore's phrase, the 'naturalistic fallacy.'" In what for Murdoch is the most important sense, this is Plato's point too: the Good transcends any particular instance associated with it. However, Moore (1922, 118) further argues that this means that "no truth about what is real can have any logical bearing upon the answer to this question."

This is often interpreted as a straightforward example of a distinction between facts about the world and values, but it is much more than this. To rephrase Hepburn's point, we might say that the attempt to define the Good in worldly terms involves a form of closure (a supposedly complete answer) that would belie the continual ethical questioning made possible by the way we use the term *Good*. It is precisely the possibility, indeed necessity, of this continual questioning—*Is that good enough? Is it really good?* and so on—that is the mark of ethics. In Levinas's (1991) terms, ethics (the Good) is a relation of infinity (openness) rather than totality (closure). As Levinas notes, the concept of infinity is precisely what is required to illustrate this ethical relation, since the ideatum of infinity always exceeds the idea of infinity, just as the Good always transcends any instance of it or attempt to define (totalize) it. What the Good referred to in such questioning *is* (what in Moore's terms Good means) seems impossible to articulate or define in its fullness despite its obvious importance in our lives. And so, as Murdoch (1970, 3) remarks, faced with this conundrum, Moore came to argue "that good was a supersensible reality, that it was a mysterious quality, unrepresentable and indefinable."

There is then more than an echo of Plato here in Moore, which is why this understanding resonates with Murdoch. It also fits with Levinas's (2004) understanding of ethics as concern for an unrepresentable and indefinable Other that is "beyond being." Levinas (19) explicitly recognizes similarities between Plato's and his own ethics: "The beyond being, *being's other,* or the *otherwise than being* . . . here expressed as infinity, has been recognized as the Good by Plato. It matters little that Plato made of it an idea and a light source." That said, as Murdoch (1970, 3) makes clear, Moore was still "a 'naturalist' in that he took goodness to be a real constituent of the world." Murdoch, too, specifically refers to ethics as immanent and incarnate (see Widdows

2005, 74). Nevertheless, Moore's talk of a supersensible reality sounds metaphysical in the stronger mystical (or in Sullivan's somewhat misleading terminology, "objective") sense. Consequently, this aspect of Moore's solution did not sit easily with the dominant strands of commonsense philosophy and scientific materialism that emerged with analytic and ordinary language approaches (approaches that, ironically, Moore himself was partly responsible for inaugurating). For this reason, Murdoch argues, many modern philosophers abandoned the very attempt to speak of the Good as such. She further traces this philosophical abandonment back to a particular, and somewhat partial, reading of Wittgenstein, who in turn offered a partial take on Moore's position.

Understanding the role Murdoch gives Moore and Wittgenstein is important because it helps indicate the very different form taken by her own (and, from her perspective, Plato's philosophy) summed up in the title of her *Metaphysics As a Guide to Morals* (Murdoch 1992). Wittgenstein's (1990) *Tractatus Logico-Philosophicus* makes a famous distinction between the world, defined in terms of descriptions of states of affairs (facts), and metaphysics, about which, he claims, nothing meaningful can be said. For the early Wittgenstein of the *Tractatus,* since ethics as such was, as Moore suggested, "unrepresentable and indefinable," it was simply ineffable, metaphysical (in a strong sense), and therefore, on his understanding, not a matter for philosophy.

This doesn't mean that Murdoch believed Wittgenstein himself thought ethics unimportant but that he considered that philosophy had nothing to say here and, as the *Tractatus* famously concludes, "Whereof one cannot speak, thereof one must remain silent" (Wittgenstein 1990, 189 #7). As Widdows (2005, 50) points out, "Murdoch believes that Wittgenstein had (at least in his early work) a strong sense of moral value," as evidenced by her quoting Wittgenstein's 1919 letter to Ficker. Here Wittgenstein (in Murdoch 1992, 29) claims "the book's [*Tractatus's*] point is an ethical one. . . . My work consists of two parts: the one presented here plus all that I have not written. And it is precisely the second part that is the important one." For Wittgenstein, the problem is that discussions of ethical values run up against the limits of linguistic expression. "In ethics we are always making the attempt to say something that cannot be said, something that does not and never will touch the essence of the matter. It is *a priori* certain that whatever definition of the good may be given—it will always be merely a misunderstanding to say that the essential thing, that what is really meant, corresponds to what is expressed (Moore)" (Wittgenstein in Murdoch 1992, 29). As Johnston (1989, 76) puts it, Wittgenstein concludes that ethics per se "involves an attempt to say the unsayable."

Now Murdoch agrees entirely with this, and therefore with Moore and Wittgenstein insofar as this is a description of the difficulties, indeed impossibility, of defining the essence of ethics (the Good). Levinas too would concur, as would Plato, on Murdoch's view. However, Murdoch, like Levinas and Plato, disagrees strongly that this means that philosophy has nothing to say (and in this sense she remains closer to Moore).[4] The inevitable result of Wittgenstein's silence was, as Murdoch (1999, 55–79) points out, that acolytes of his early philosophy accepted and further developed this extreme version of the fact–value distinction, the search for linguistic closure, and an excessively narrow view of philosophy.

Murdoch wants to maintain the importance of this metaphysical ideal for the conduct of ethics and to defend Moore against those critics who regard what is indefinable as meaningless, as an empty concept. We constantly employ the idea of Good in our ethical debates despite that we cannot point to it, quantify it, or locate it definitively. Indeed, she argues, we need this concept: first, because an ethical ideal, a perfect (if ultimately unattainable) form of the Good, informs our understandings of how to relate to others unselfishly, and second, because she believes a unitary concept of the Good necessarily emerges as we make connections between different instances of good actions, of justice, benevolence, tolerance, and so on. We should not then, she argues, abandon the idea of the Good as a unitary (but indefinable) concept because, as we reflect on the richness and diversity of ethical language, we recognize that all these terms are interconnected even though what ethics might require always exceeds any possible list of examples or any given definition. Perhaps, then, following these reflections, we might propose an understanding of ethics as drawing upon a utopian idea of the Good, which is ultimately incomprehensible as a conceptually defined totality. Ethics is expressed in a necessarily imperfect world as the wisdom that offers to shelter and conserve the never fully represented excess of others' being, their earthly existence such as they are.

Interestingly, what concerns Murdoch, as despite their immeasurable differences it did Bataille, is this recognition of the existence of an indefinable excess that always escapes the totalizing claims of the dialectic—a dialectic that, whether Platonic (rational) or Hegelian (historicist), is itself driven by a desire for the inclusion of everything that is potentially meaningful. Everything beyond this remains literally meaningless—as meaningless as metaphysics to Wittgenstein's naturalistic followers. But Murdoch and Bataille understand philosophy differently, as a paradoxical yet ongoing attempt to say something that is ultimately unsayable, which cannot be captured (represented) fully within language. Although many philosophers have fallen prey to the totalizing (all-consuming) desire to make everything fully presentable, philosophy has to be more than this impossible attempt to bring everything into systems of representation. Rather, philosophy must begin and end with that wisdom which, paradoxically, knows that we cannot know everything, the wisdom to know that what there is, what exists, does not passively await our representation of it and that the knowledge relation, even as it reveals the world to us, is always also one of concealing incompleteness. This is not an abstruse mysticism; it is a sensible, philosophical realism.[5]

It is surely pertinent to ask whether Plato was already aware of the paradoxical limits of the dialectic in revealing the Good and the truth. Had he learned from Socrates what Socrates had traveled widely to discover, that however knowledgeable one might be, true wisdom lies in acknowledging one's ignorance? "A knowledge of our own ignorance is what human wisdom is" (Gadamer 1985, 185). When Socrates claims that the oracle at Delphi judged him the wisest of men because only he was aware of the depths of his ignorance, this was not just a form of false modesty but an integral part of his understanding that self-reflexive participation in the dialectic actually reveals such ignorance to both parties. As Gadamer (1980, 93–123) argues, the Socratic dialectic is not a form of proof, it is not a matter of *compelling* agreement. The Socratic elenchus expresses its own inadequacies even as it reveals, as it so often does, only aporia, those gaps in knowledge that are the wellsprings of argument but which also suggest the void beneath every thinker's argumentation. In philosophy, truths can often only be suggested from afar, intuited, since they are not amenable to the conclusive proofs of any logic that is itself dependent on the fullness of symbolic representation. Gadamer emphasizes the paradoxes, the aporia, the utopian aspects of the *Republic,* and the ironic (although often embittered) humor at play within Plato's discourses: these all pay heed to the wisdom necessary to recognize one's own ultimate ignorance.[6]

This is also why philosophy, in its etymological sense, is conceived of as an ethical concern, a love for wisdom *(philosophia)* about life in the face of death. For, as Bataille points out, death is the ultimate limit of meaning, the resolution of life, that which threatens to make *everything meaningful meaningless* and yet that which, if we are to be truthful to ourselves, we must face even though we can never ourselves experience it.[7] This is also why, as we have already seen, for Bataille, history is never a completely closed system, why a residue remains even at the supposed end of history, at the completion of Hegel's dialectic, when even philosophy must cease. This incompleteness might be explicated in terms of Bataille's excessive Hegelianism—the ways in which Bataille pushes Hegel's system beyond its self-imposed limits as a closed system by refusing to regard the process of historical over-coming *(Aufhebung)* as one in which each stage is successively and completely conserved and negated within the next.[8]

The understanding of ethics (the Good) and of truth as a continual questioning suspended above an abyss of infinite ignorance (which these more secular if less conventional readings of Plato's emphasize) would hardly provide a compelling argument that philosopher-kings should wield absolute or sovereign powers! On this reading, Plato's *Republic* is far from having provided a solution to the vicissitudes of life, the uncertainties of politics, or the ambiguities of ethics; rather they are writ large in the very idea of the Good and in the failure of that idea to grasp its ideatum. The claim that the philosopher-kings can dialectically attain knowledge of the Good as a basis for absolute political rule could no longer be understood in terms of their being able to access metaphysical certainties about the way the world is; quite the contrary. The dialectic would not reveal timeless, overarching ideals and principles, rather, like Socrates's own subversive, anarchic, and always inconclusive practices, it suggests the necessary openness of all philosophical, ethical, and political debate, the essential impossibility of their (en)closure, their roles in the immanent critique of metaphysics' totalizing tendencies.

And so, just as Murdoch, Moore, and Levinas, each in their own way (and through their own readings of Plato), regard ethics as a relation of infinity, a continual questioning of the adequacy of our responses to other beings, simply thinking of ethics in this way begins to reveal inadequacies in dominant understandings of almost every aspect of contemporary life—the wider patterns of potentially damaging but systematically imposed world domination whose origins and ecological effects, as related in the *Critias*, even predate Plato. It reveals, for example, the errors underlying the dominant tendency to dismiss that which lies unsaid, unrepresented, beneath the surface of language, reason, and history; to take the system of human representations for the whole and act into the world on this basis. In this way, information is mistaken for wisdom, economic price for an entity's value, the scientific calculus of risk for real dangers, political representation for real politics. In a way, this lack, this unachievable (and ultimately unethical) desire for completeness, this absence of wisdom, is the *real* source of so many of our ecological problems (Sandilands 1999, chap. 8; and see later in this chapter).

Worldly (In)Difference and Ecological Ethics

> A self-directed enjoyment of nature seems to me to be something forced. More naturally, as well as more properly, we take a self-forgetful pleasure in the sheer alien pointless independent existence of animals, birds, stones, trees. "Not how the world is, but that it is, is the mystical."
>
> —Iris Murdoch, *The Sovereignty of Good*

uncertanty

Might Murdoch's quasi-Platonic understanding of ethics, of the *Sovereignty of Good,* begin to offer alternatives to anthropological despotism and/or pastoral stewardship, alternatives more amenable to expressing ecological concerns and responsibilities? There are clearly at least two aspects to this question. The first concerns the extent to which the Good is understood in terms of the limits placed on ethics by the anthropological machine and its desire to eradicate all ambiguities concerning the differential status of human and nonhuman beings, to confine, for example, the animal to the realm of things. The second (as already mentioned) concerns the nature of sovereignty itself and whether an understanding of ethics *as such* could be used as a principle of political legitimacy in terms of a secularized form of Earthly stewardship. Despite her book's title, the reading of Murdoch's work presented here casts grave doubt on this, suggesting that the very idea of the sovereignty of Good is contradictory.

To turn then in this section to the first aspect: the relation between a potentially ecologically oriented ethics—the Good (understood in Murdoch's sense)—and the anthropological machine. Any ecological ethics has to struggle against the forgetful tendencies of a dominant culture that has come to regard all of nature instrumentally, that is, as no more than a resource the meaning and value of which lies only in its potential to be transformed and used in the service of humanity. Giving nature ethical consideration requires, in this sense, a form of an-amnesis, a remembering of the troubling ambiguities that (as Lascaux illustrates) have, since the beginning, surrounded the self-defined sta-tus of humanity in its relation to the natural world. The culture's denial or evasion of wider ethical responsibilities, this putting nature out of mind, is dependent on the ideological successes of the anthropological machine, now most frequently present under the auspices of Homo economicus, in reducing nonhuman beings to mere objects.

From this narrowly anthropological perspective, the only feasible and rational approach toward valuing nonhuman nature is to espouse an enlightened self-interest, whether couched in terms of an appeal to human individuals, human communities, and/or humanity as a whole (for example, the purported interests of the human species in its self-preservation). The ethical differences between these levels of appeal are elided or (more rarely) exaggerated as political sensibili-ties require. For example, the so-called Brundtland Report (WCED, 1987), the founding document of sustainable development, assumes both a common self-interest in averting ecological damage and a gen-eral (ethical) concern for future generations of humans while glossing over the self-interested individualism that is the ideological lynchpin of the environmentally destructive economic system of global capital-ism.[9] Often portrayed as a model of ethically responsible stewardship, it should not be surprising that Brundtland uncritically adopts the bio-political notion of "household management" that has already become a focus of this critique, concerning itself only with managing nature as a human resource.

Murdoch's *Sovereignty of Good* offers an ethical critique of indi-vidual self-interestedness that might also be pertinent when it comes to the imposition of these broader human-centered patterns on the natural world. This is partly because her target is not just the delib-erate individual selfishness characteristic of Homo economicus but self-centeredness in a much wider sense of the sovereign individual. The "self-directed enjoyment" quoted at the beginning of this section would also include all those tendencies to envisage others (human or nonhuman) as simply being "there for," as fashioned after the model of, or revolving around, that more or less idealized human being that individuals take themselves to exemplify. It refers to the imposition of their self-concernedness on the wider world, and it is these self-directed concerns that Murdoch thinks ethics dissolves and exceeds, perhaps especially in the presence of alien, more-than-human others.

Encounters with animals, birds, stones, trees, Murdoch (1970, 84) suggests, offer occasions for "unselfing," a quality of experience that attends "to nature in order to clear our minds of selfish care." "I am looking out of my window in an anxious and resentful state of mind, oblivious to my surroundings, brooding perhaps on some damage done to my prestige. Then suddenly I observe a hovering kestrel. In a moment everything is altered. The brooding self with its hurt vanity has disappeared. There is nothing but the kestrel. And when I return to thinking of the other matter it seems less important." Murdoch is explicitly *not* describing a feeling of romantic self-exaltation here, but even so, the predominant tendency might still be to try to frame her interests in the edifying and inspirational pleasures afforded by nature as somehow self-centered in a more tenuous sense. From this self-reductive perspective, such experiences would also fall short of any ecological ethics insofar as they emphasize the value in being relieved of troubling, inwardly directed cares rather than invoke an outwardly directed, worldly concern for nonhuman others like the kestrel.

Insofar as Murdoch was not, in any sense, trying to sketch an environmental ethic, the potential existence of such a gap would hardly be surprising. However, since her interests lie in ethics, not individualistic well-being, it is important to take her claim of unselfing seriously, to resist the tendency to recuperate and resituate her thoughts within those self-referential frames that reduce all else (including the kestrel) to beings of merely instrumental value, however enlightened the purposes they might serve. This would indeed be a forced interpretation and one just as damaging to Murdoch's own immediate concerns as to any potential ecological ethics. The important aspect of nature for Murdoch (1970, 85–86) is that (like good art) it "offers a perfection of form which invites unpossessive contemplation and resists absorption into the selfish dream life of the consciousness."[10] This clearing of the mind from self-oriented concerns might be better understood as a condition of ethics per se, a feature of an enlarged field of ethical sensibilities that "transcends selfish and obsessive limitations of personality"(87). Because of this, this clearing is also, argues Murdoch, an attempt to see the world as it really is, stripped of the self-centered illusions we compose to console our human psyches and therefore to return to the originally quoted passage, "alien" (estranged), ultimately "pointless" (nonteleological), and "independent" (existing in its own right, owing nothing to ourselves or humanity).

This then, according to Murdoch, is the "nature of the world" that humans inhabit: it is both how the world is and what nonhuman nature reveals to us. As such, it might appear unfertile ground for any kind of ethics, human or ecological. But there are two crucial issues here worth examining in more detail. First, for Murdoch, recognizing this indifferent environment in no way forces us to accept those dominant modernist cultural and philosophical forms that portray the human self as the sole remaining source and measure of ethical values.[11] Quite the opposite: self-obsession, whether it takes the form of the rational Puritanism of Kant where values "collapse into the human will" (Murdoch 1970, 80) or the "romantic self-indulgence" of someone like Kierkegaard, who emphasized "suffering freedom" (82), simply leads us back into illusory self-containment. It hides from us the reality of the world's indifference by again making our self-referential anthropocentric concerns central; it replaces a world-transcending God made in humanity's image with a simulacrum of humanity itself, monotheistic religion with a monolithic variety of humanism (yet another form of the anthropological machine).

Second, the task of ethics necessarily involves coming to see the world as it really is, and this is just as true for any human ethics as it would be for an environmental ethics. By this Murdoch does not mean to simply argue that some kind of detached epistemic objectivity, the attempted removal of personal prejudices and selfish desires, is necessary to avoid clouding our ethical judgments. She means that attending selflessly to "how the world is" is intimately connected to, indeed inseparable from, attending to "how we should respond to the world." That is, to succeed in acting ethically toward someone would entail having succeeded to some extent in seeing them as they really are. Why? Because this, despite all its difficulties, is what ethics requires: that one regards and responds to others in the light of who they are for themselves, not who one assumes, fantasizes, or would prefer them to be. And here we might think of what it means to truly love someone and how dependent this is on (not stifling) their ability to act in ways that surprise us and take us away from our everyday self-centered interests.

From Murdoch's perspective, ethics is, at its very heart, an exercise in attempting to see and respond to the world as it is; a project that requires a suspension or clearing of our self-referential obsessions, the distorting influences of that self-regard which always tends to reconstitute others as somehow being like us, revolving around us, or suiting our interests. Ethics is an awareness of others' differences and independence from ourselves. A good person must "know certain things about his surroundings, most obviously the existence of other people and their claims" (1970, 59). There are many ways in which, in failing to attend to the reality of such differences and to the world's indifference to us (its independent existence), we easily return "surreptitiously to the self with consolations of self-pity, resentment, fantasy, and despair" (91). However, to be ethical means resisting the temptation to relate to the world as if we were directors of our own personal Hollywood films, forcing others to play prescribed roles or reducing them to depthless characters in pursuit of our predetermined but, from others' perspectives, fictional ends.

This understanding of ethics is, as already intimated, actually close (though by no means identical) to that suggested by Emmanuel Levinas. Levinas's extensive writings might be considered as attempts to overcome the recurring tendency to center our worldly (and philosophical) understandings on the "egocentric monism" (Peperzak 1993, 19) of the human self. This egoistic self, in its pursuit of its own "closure and contentment" (Lingis 2004, xxii), its "self-sufficiency," strives to reduce the world to one where everything turns on and reflects the interests, concerns, and form of the "I" who beholds it. This can provide an illusory comfort: "I am at home with myself in the world because it offers itself to or resists possession" (Levinas 1991, 38), but this self-centeredness, this notion of the sovereign individual, is the antithesis of an ethical relation; it relates to the world on the basis of self-possession, reducing its otherness, its alterity, to an economy of the Same—a system of relations based on self-identified desires. Ethics is, however, not a relation of possession at all.

Again, it is worth remembering that although Murdoch's and Levinas's arguments may sound philosophically abstract, we do indeed inhabit a social world in the thrall of an economically reductive model of self-centered individualism—Homo economicus. The capitalist economy, with its reduction of everything to a resource whose purpose is to fulfill our self-identified needs might then, from a Levinasian perspective, be thought of as a particular instance of the fundamental structure of the economy of the Same (although this seems ahistorical insofar as it fails to consider the actual role of capitalism in producing the particular form of a possessive individualism taken as timelessly given by both Levinas and Murdoch). Ethics, from this neoclassical and neoliberal perspective, is something that runs counter to the realism of the market economy and to realpolitik. From a narrowly pragmatic perspective, ethicists, especially environmental ethicists, are simply living in a dream world, one they should abandon in order to focus on the kind of solutions that could work with society's self-interested concerns.

Murdoch (1970, 78), like neoclassical economists, believes that "human beings are naturally selfish" (Levinas, too, recognizes a pervasive economy of needs), but she then turns current opinion about what this entails on its head. For her, this is not a situation we should simply accept but a problem to be overcome precisely because it leads to our constructing self-centered fantasies that bear little resemblance to reality: "fantasy (self) can prevent us seeing a blade of grass just as it can prevent us seeing another person" (78). And to see a blade of grass (ethically) is precisely not to see it in terms of its instrumental use for us. For Murdoch, then, it is not ethics that is guilty of idealizing the world: far from it, ethics is the ultimate form of realism. One might argue that the all-pervasive forms of economic and political realism are actually guilty of propagating and pandering to the world-distorting influences of self-centered concerns.[12] For this reason, if for no other, such political realism goes hand in glove with worldly inattention and, we might add, its almost inevitably deleterious social and environmental consequences.

By contrast, attending to the world as it presents itself to a loving rather than self-interested gaze (which is what Murdoch [1970, 34] means by "attention," a term she takes from Simone Weil) is both a coming to see the world as it really is and a route to understanding what ethics is. The natural "perfection of form," which we glimpse in our wonderment at the kestrel's flight, provides an inkling, in its selfless apprehension, of the perfect forms, the idea(l)s of the True and the Good (reality and ethics). Through "unpossessive contemplation" of the radical otherness of nature, we open the possibility that we might realize, in both the sense of making real (however incompletely) and recognizing (however dimly) something of the idea(l)s that inform, yet are never wholly contained within, any particular instance of worldly attention. That is, we indirectly get an idea of what Good is, what it is that informs, say a particular act, in such a way that it becomes ethical.

Here, too, there is more than a passing resemblance between Murdoch and Levinas, though for him it is not through contemplating nature that we come face to face with the Other, this "Stranger who disturbs the being at home with oneself" (Levinas 1991, 39). For Levinas, it is in witnessing the face of another human that we encounter a force that contests and unpossesses the ego, that exposes us to an alterity, a difference, which cannot be assimilated to an economy of the Same. The human face, like the kestrel's flight for Murdoch, is an epiphany that bears its own significance. It is more than a visage, it also offers a *sur*face, an opening onto that beyond, that which transcends the phenomenology of its appearance: that is, in our being called to the task of ethics, we glimpse the (in)different (strange, pointless, independent) reality of the Other.[13] This reality, who the Other really is, always exceeds our experience of them: they are so much more than what we see before us, than our knowledge of them, than any words that seek to define (and thereby contain and limit) their identity. And yet it is this reality, and their alterity, that their faces reveal, however fleetingly and incompletely. Levinas then uses the term *face* both literally in terms of the face of *an*other individual, someone we might meet, and to express the irreducible alterity of *the* Other, the nonreducible reality behind appearances that confronts us in any such ethical encounter as an instance of otherness as such. In Peperzak's (1993, 64) terms: "'Face' is the word Levinas chooses to indicate the alterity of the Other forbidding me to exercise my narcissistic violence."

Perhaps it is now possible to begin to think the relation between ethics and worldly (in)difference in something like Murdoch's terms. Her concern, like that of Levinas, is to say something revealing about ethics as such, to relate something of its worldly vitality, without treating it matter-of-factly. The Good (ethics) is ideal (utopian, one might say) but only in the way that it expresses a perfect relation to others, one of infinite openness rather than totalizing closure. This is *not* the espousal of a form of philosophical idealism. "Goodness is an idea, an ideal, yet it is also evidently and actively incarnate all around us. . . ."

(Murdoch 1992, 478). It denotes a *de*limitation of the human condition, a transcendent reality and a worldly (im)possibility for a human existence that can move toward, but never fully attain, such self-effacement in its pure form. (Derrida [1995b] argues a similar point in discussing the ethics of the "gift"; see also Smith [2005a] and later in this chapter.) Murdoch (1970, 93) sums up this situation: "The self, the place where we live, is a place of illusion. Goodness is connected with the attempt to see the unself, to see and respond to the real world in the light of virtuous consciousness. This is the non-metaphysical meaning of the idea of transcendence. . . . 'Good is a transcendent reality' means that virtue is the attempt to pierce the veil of selfish consciousness and *join the world as it really is*. It is an empirical fact about human nature that this attempt cannot be entirely successful."

Leaving aside for the moment the question of how far Murdoch's and Levinas's understandings might be compatible, or at odds, with claims about the ethicopolitical stewardship of nature (questions that, as the next section illustrates, are closely connected with the manner in which their metaphysics is thought of as providing a "*guide* for morals"), it is still necessary to ask what it means to "join the world as it really is" and how this might relate to a potential ecological ethics. In other words, how far might such approaches be capable of recognizing the ethical import of nonhuman others given that both Murdoch and Levinas speak of the other as a *human* being? In Levinas's terms, the Other (*Autrui* often, but not always consistently, capitalized) is exclusively and explicitly so, as, for example, with regard to the face-to-face encounter.

Certainly, if such an ethics can be understood as being relevant to the more-than-human world, it offers the possibility of paying concerned attention to patterns of difference in nature without reducing these differences to representational codes (taxonomies) and systems (axiologies) that might claim to, but cannot, capture essential moral distinctions between categories of beings (Smith 2001a). Such an ethics would be a much more suitable response to a natural world that is alien, purposeless, and independent of human interests. Animals, birds, stones, trees really are alien in the sense that they are other than human, that they exhibit radically different and sometimes extraordinarily strange ways of being-in-the-world. Humanistic approaches, indebted to the anthropological machine, tend to emphasize and use these differences as reasons for excluding such things from moral consideration. They are not like our-human-selves, and so, they argue, in their anthropocentric self-obsessed ways, can consequently be of no ethical (as opposed to instrumental) interest to us.

The unfortunate response of environmental ethics to such claims has often been to try to minimize differences and find essential similarities or common purpose or to establish mutual dependencies by extending these same self-centered patterns (Taylor 1986; Attfield 1991). Certain aspects of the environment are deemed morally considerable because they share some supposedly key aspect of human selfhood that makes them as "intrinsically" valuable as our*selves,* for example, as subjects-of-a-life. Our self-concern becomes the basis for a (supposedly) ethical concern for those others deemed sufficiently like us. An alternative, more expansive strategy, which still retains this same self-centered form, is to suggest that the whole of nature might be deemed valuable insofar as it is reconceptualized (via, for example, ecology, quantum physics, or non-Western metaphysics) as part of our extended selves (see, for example, Callicott's [1985] early work). Some even combine both strategies, for example, by espousing a form of "contemporary panpsychism" whereby the universe is reenvisaged as a "self-realizing system," which "possesses reflexivity and to this extent . . . is imbued with a subjectival dimension" (Mathews 2003, 74).[14]

However, in adopting these strategies, these purportedly biocentric approaches change the content but retain the form, the same anthropocentrically self-obsessed locus, of the dominant ethical field (Smith 2001a). These forms of axiological extensionism, while often well intentioned, are not only philosophically artificial (constructed largely in order to justify certain already predetermined ends) and ecologically impractical but also tend to replicate, rather than fundamentally challenge, the presuppositions of the anthropological machine. For all their egalitarian rhetoric, they tend to ethically favor those things most like, or closest to, that defined as properly human. The real differences that an alien nature presents are overlooked and human alienation fantasized away.[15] By contrast, Murdoch and Levinas can be understood as arguing that ethics exists as a non-self-centered response to the recognition of such alienation from the world and from others. Indeed, there is no real ethics without recognizing such differences. An ecological difference ethics thus potentially offers a radical alternative to all attempts to enclose the nonhuman in an economy of the Same.

If this explains why a difference ethics might be important, it also raises the question of the metaphysical limits of Murdoch's and Levinas's own humanism, their own indebtedness to the anthropological machine—that is, the degree to which their philosophies fail to consider our relations to nonhuman others and the extent to which this failure is an inherent feature of, and expressed by, their understanding of ethics as such. Levinas (2003) explicitly develops a *Humanism of the Other*, where we become properly (ethically) human, demarcating ourselves from our natural (animal) selfishness, through the call to responsibility made immediate in another human "face." This form of humanism may be radically different from those based on self-possessive or reflexive individualism—it puts the ethical relation to the Other first, above all else—but only the human Other. It may claim to be precultural and socially transcendent insofar as "the nudity of the face is a stripping with no cultural ornament—an absolution" (Levinas 2003, 32), but it is so, ironically, only because it treats members of the human species "fellow man [sic]" (7) as similarly special, as exemplifying a (uniquely ethically important) difference. The paradox is that the call to responsibility comes into being precisely because "humanity is not a genre like animality" (6), because the transcendent differences of another individual human cannot be denied, yet this still treats humans as a genre. And so, although the ethical relation is nonreciprocal in the sense that it does not imply a return—some reciprocal recognition of or benefit to—the individual valuer, it nevertheless implies a return in the sense of a metaphysical indebtedness to the community of humans individually and collectively composed by such ethical relations. "*We* recognize ethics . . . this *obsession* by the other man" (6). For Levinas, *I, We,* and *the Other* are all constituted through an ethical relation that is ironically defined as an obsession with only those differences that lie beyond the phenomenology of properly human beings.

Levinas has, of course, been criticized for the anthropocentrism of his metaphysical assumptions. Most famously, this self-imposed limitation on his thought, the closure implicitly underlying his otherwise nontotalizing form of humanism, was the subject of commentary by Derrida (1992, 2008). And as David Wood (1999, 32) notes in his discussion of this commentary, the "question of the other animal is . . . an exemplary case because once we have seen through our self-serving, anthropocentric thinking about other animals, we are and should be left wholly disarmed, ill-equipped to calculate our proper response. It is exemplary because the other animal is the Other *par excellence,* the being who or which exceeds my concepts, my grasp, etc." And, we might add ecologically, if this is so for animals, then it is even more so for trees or stones.[16] The fact that Levinas does not recognize this suggests strongly that there is indeed a mystical (in the strong sense) aspect of his metaphysics where humans are concerned, a special kind of "beyond being" that applies only to humans. The rootedness of his philosophy in Judaic religious traditions with the "fundamental emphasis it places on inner [human] life" (Wyschogrod 2000, 178) would support this reading. But, for these reasons, any attempt to develop a Levinasian formulation of a nonhuman difference ethics must include a critical (secularizing) appraisal of the metaphysical assumptions that ground his philosophy too.

Extending a Levinasian model of the face-to-face encounter to other nonhuman faces—kestrels, for example, or perhaps ecologically even to rock "faces" or landscapes—is by no means as simple as it might seem. Levinas, as Diehm (2000, 53) notes, is resistant to, or at best ambiguous about, such moves. And while Diehm argues that "I do not think that Levinas's ethical phenomenology of the face precludes other-than-human faces" (56), his attempt to propose such an extension seems plausible only because he downplays the metaphysical aspects of Levinas's ethics. "I want to argue," says Diehm, "that when Levinas says 'face' what he really means is 'body' and that it is on the basis of this understanding that we can speak of the face of the other than human" (54). But, as noted previously, Levinas's understanding of face has an integral, indeed crucially, metaphysical aspect that, from his perspective, might well preclude such a move (see later in this chapter).

An understanding of metaphysics is no less crucial where Murdoch is concerned. However, while her humanism is, like Levinas's, critical of any instance of that "cult of personality" (Murdoch 1999, 275) that ethics seeks to overcome, it is not limited in the same way to encounters with the human Other. To be sure, her main concern is with "respect for the [other] individual person as such" (275), but not only are animals, birds, stones, and trees openings into the unselfing of ethics, they can, it is sometimes suggested, also elicit calls for responsibility: "We cease to be [selfish] in order to attend to the existence of something else, a natural object, a person in need" (Murdoch 1970, 59). There is no a priori reason why Murdoch's metaphysics need be as anthropocentrically limited as Levinas's. Indeed, the mysticism present in her metaphysics is much more worldly than Levinas's partly because it is less tied to specific religious beliefs about human specialness (Bayley [1999, 121] describes her as "religious without religion") and more about a general recognition of the importance of those aspects of life that cannot be encompassed in the reductive, representational world of Wittgensteinian "facts." One might say that Murdoch offers a perspective that has begun to be ecologically "secularized."

Clearly an ecological ethics has to be paradigmatically worldly, at least in terms of its concerns. And here it is important to stress again that the metaphysical emphasis that characterizes Murdoch's and Levinas's difference ethics is not, as some critics have suggested, otherworldly. The point of speaking about transcendence and infinity is not to locate ethical values in some supernatural realm. It is to try to say something about what Wittgenstein thought unsayable, to gain an insight into a nonreductive understanding of that which cannot by its nature be comprehended in its totality—namely, ethics as such. Once understood in this way, ethics requires that we recognize that there is always more to others and to the world than we can ever see or know and that significant others are and should be valued as beings that are different, independent, and irreducible to our purposes.

Of course, to pay attention only to that "beyond being," that which cannot be comprehended, risks falling into obsession with a transcendent realm at the cost of embodied individuals. But despite Alford's (2002, 37)[17] claim that "Levinas was never interested in the concrete reality of the other person, whose fleshy reality can only get in the way of transcendence," this is not what Levinas or Murdoch aspire to. Murdoch's utopianism regards Good as an idea(l) toward which we might make faltering steps by trying to understand others, however difficult this might be: "one aspect of respecting something is being interested in it enough to try to understand it" (Murdoch 1999, 275). In this sense, there is a hermeneutic aspect to Murdoch's ethics that is somewhat lacking in Levinas's moral absolutism.

Levinas gets caught up in the extraordinary metaphysical subtleties necessary to unpack his ethics philosophically and in the performative difficulties in "saying" such things without their being thereby fixed in conceptual aspic. And this is perhaps another reason why Murdoch's approach seems more down to earth, since she wanted to directly address, however critically, the commonsense philosophy of her peers rather than engage in more abstruse metaphysical poetics or mechanics. She, for example, says relatively little about the actual process and repercussions of unselfing, whereas Levinas focuses intensely on this issue. For this reason, much of what he says might usefully inform Murdoch's views, while much else, due to his specific metaphysical predilections, might serve only to narrow her position anthropocentrically. Again, any potential ecological (difference) ethics needs to carefully consider such issues.

Murdoch is also more down to earth in another sense crucial for any ecological ethics: she directly links the transcendent nature of ethics with nature's own necessary transcendence of its human-centered appearances. For Murdoch (1970, 47), "Good is indefinable not for reasons offered by Moore's successors . . . but because of the infinite difficulty of the task of apprehending a magnetic but inexhaustible reality." Ethics is, in this sense, a task that emerges through recognizing the infinitely complex composition of the world. "If apprehension of good is apprehension of the individual and the real, then good partakes of the infinite elusive character of reality" (42). In other words, instead of focusing on the strongly mystical infinity of the human face, Murdoch continually leads us back to the links between ethics and nature itself, which exceed, surpass, resist, and go beyond all attempts to capture them in formulaic generalizations.

Such insights, of course, often inform good literature (art), including Murdoch's better novels and our actual experiences of nature, but are soon forgotten or overwritten in the constant procession of activities, conceptual schemes, and social systems directed toward accounting for (defining and enumerating) what there is in the world and its value, for example, in terms of axiologies, taxonomies, bureaucracies, markets, systems ecology—indeed, almost every mode of modern life that revolves around the need for completeness and mastery over society and nature. Humanistic models of ethics also tend to buy into this approach, but in trying to fold ethical experience back into contemporary structures, they lose sight of ethics as such. Murdoch (1970, 42–43) quite explicitly differentiates her own understanding of ethics from such philosophy: "I have several times indicated that the image which I am offering should be thought of as a general metaphysical background to morals and not as a formula which can be illuminatingly introduced into every kind of moral act. There exists, so far as I know, no formula of the latter kind." Those who want to make ethics useful, to instrumentalize it, make it pay its way in the modern world, or apply it formulaically to ecological concerns will not like this at all. But Murdoch's point is precisely that ethics ceases to be ethics if understood or used in this way. Interestingly, the heartfelt desire for an uncompromising ideal of ethics, this constant questioning of what the world is like (is how we relate to nature Good?) is the animus that informs radical ecological politics too.

There is clearly much more one could try to say here, but bearing in mind these metaphysical caveats, Murdoch's ethics combined with aspects of Levinas's have the potential to offer ecological ethics the breathing space it needs. It transforms our understanding of the ethical landscape in such a way that the human self is radically decentered from its position as the sole source of truth about, or value of, the world. Of course, in doing so, it raises a series of fundamental questions about how ethical relations among human selves, the world, and other diverse beings might be conceived; questions that might be approached using rather different philosophical forms and concepts than those currently predominating. But the plausibility of this decentering move is enhanced once we recognize ethics can be envisaged as a form of concerned response to the recognition of the discomforting realities of a world described in Murdoch's sense as alien, pointless, and independent. Not that there is anything automatic about the appearance of such responses: ethics does not leap fully formed to fill the world with meaning and value; it is, as Murdoch suggests, a task, and an ambiguous one at that. Our self-centered concerns are not easily and never fully dislodged, especially when the very idea that they should be seems to require the ethical equivalent of that other Copernican revolution mentioned in the Introduction.

However, if we follow Murdoch's, albeit unintentional, lead here, we can see that the justificatory onus might be shifted away from having to defend ecological values in terms of the formulaic extension of ethical systems modeled on human self-concerns. If nature is to be ethically considerable, this is not because we have to recognize something of ourselves in its operations. It is rather our attending to the (in)different realities of the natural world that opens the possibility of any ethics at all. Simple attention to the flight of the kestrel, this momentary losing of our self-obsessions, becomes an opening on ethical possibilities because it shows us, however fleetingly, how things "really" are. Momentarily cleared of self-interest, the condition of the world and the human condition appear anew, though perhaps not as our self-centered humanistic fantasies would have wanted them to be.

Sovereignty, Ethics, An-archē

> Morality is after all the great central arena of human life and the abode
> of freedom. Almost all our thoughts and actions are concerned with the
> infinitely heterogeneous business of evaluation, almost all our language is
> value language. The destruction or denial of this open texture is and has
> been (as we know) the aim of many theorists and many tyrannies. Moral
> (that is human) activity can be controlled if it is conceptually simplified.
>
> —Iris Murdoch, *Sartre: Romantic Rationalist*

The metaphysically weaker understanding of ethics described in the
previous two sections seems to offer possibilities for ecological ethics to
escape the clutches of the anthropological machine. But how might it
fare in relation to politics? In particular, can it escape the despotic aura
that, as Arendt remarks, clings to the strongly metaphysical reading
of Plato's *Republic* and its anticipation of the biopolitical model of a
pastorate under the sovereignty of Good?

As argued previously, Murdoch's reading of Plato suggests that the
notion of Good, of ethics as such, can be understood in ways that are
very far from positing the kind of supernatural and/or naturalized
metaphysics that might support an ideocratic (philosophical) despotism.
Rather, her concern is to recognize and conserve the "open texture" of
ethics and warn of the totalizing and tyrannical dangers of attempts to
define what Good actually is and then deploy these necessarily overly
simplistic definitions in place of ethic's continual questioning, its secularizing refusal to abide metaphysical absolutes. Ethics is a practice
of "deepening and complicating" our understandings (Murdoch 1970,
31), a kind of wisdom that can be aided by concepts and metaphors
but cannot be subsumed under them. (Words can "occasion" but do not
"contain" wisdom [32]). Ethics, for Murdoch, is sovereign only in the
very different sense that it originates, overrides, and draws us outside
of any particular fixed or supposedly overarching philosophical principles set in place to define what the good life should be.

The ecological power of ethics would then lie in the way it constantly overflows, exceeds, subverts, and delimits any and all decisions
about the properly human that the various iterations of the anthropological machine employs to put an end (in one way or another) to questions concerning our responsibilities to the more-than-human world.
Ethics undermines the anthropological machine from both sides of its
prospective divide. First, as Agamben (1993, 42) argues, the "fact that

must constitute the point of departure for any discourse on ethics is that there is no essence, no historical or spiritual vocation, no biological destiny that humans must enact or realize. This is the only reason why something like an ethics can exist, because it is clear that if humans were or had to be this or that substance, this or that destiny, no ethical experience would be possible." Second, on Murdoch's and Levinas's readings, ethics is always initially a matter of a relation to an Other's singular realities, to their singularity as (lovable) beings as such and not just as members of abstract categories. Rejecting a fixed human essence and emphasizing the singularity of ethical experiences may initially make Murdoch sound like an adherent of Sartrean existentialism, although she is far from being so.[18] Rather, this singularity has more in common with Agamben's (1993) notion of "whatever" being *(quodlibet ens)*, being such that it always matters to (concerns) us. The point is that "love is never directed to this or that property of the loved one (being blond, being small, being tender, being lame), but neither does it neglect these properties in favor of an insipid generality (universal love): The lover wants the loved one *with all its predicates,* its being such as it is" (Agamben 1993, 2).

Murdoch's suspicion concerning attempts to define ethics in terms of abstract categories of being(s) is shared by Arendt. As Judith Butler notes, in an exchange of letters following the publication of her account of *Eichmann in Jerusalem* (Arendt 1994 [1963]), Gershom Scholem accused Arendt of lacking a traditional "love of the Jewish people" *(Ahabath Israel).* Her reply: "I have never in my life 'loved' any people or collective—neither the German people, nor the French, nor the American, nor the working class or anything of that sort. I indeed love 'only' my friends and the only kind of love I know of and believe in is the love of persons. Secondly, this 'love of the Jews' would appear to me, since I am myself Jewish, as something rather suspect. I cannot love myself or anything which I know is part and parcel of my own person" (Arendt, letter to Gershom Scholem in Butler 2007, 26). Arendt thus emphasizes how ethics is a relation to others, not the self, and that Scholem's idea of love would have to depend on defining who is and is not a Jew and therefore worthy of this love. In Agamben's terms, any such definition employs the same modus operandi as the anthropological machine by creating abstract ethicopolitical inclusions and exclusions concerning who is, or is not, counted as properly human/Jewish. And so, as Butler (2007, 27) remarks, "when Arendt refuses to love 'the

Jewish people,' she is refusing to form an attachment to an abstraction that has supplied the premise and alibi for anti-semitism."

In other words, these disparate theorists all agree that ethics as such is not reducible to, or constituted by, the application of universal principles to abstract categories of beings. Ethics is rather a mode of being in relation to singularly significant Others. But how, then, can such an ethics possibly relate to the realm of politics (whether ecological or not)? Before tracing the various possibilities, and in particular those opened by Arendt's own work, it has to be said that Murdoch was less attentive than she might have been of the political implications of declaring ethics a sovereign power (albeit one that operates through continually questioning all other claims to absolute authority).

Murdoch often speaks of the way in which ethics draws us out of our self-interested predilections and how it exemplifies our attempt to see the world as it really is, in terms of both transcendence and hierarchy. Ultimately, of course, the *"authority* of morals is the authority of truth, that is of reality"* (Murdoch 1999, 374, my emphasis). However, just as Plato "implies that there is a hierarchy of forms" (Murdoch 1970, 95), so when Murdoch speaks of Good as "sovereign over other concepts" (102) and as facilitating a unitary ordering of the relations between different virtues, there is a risk that this will be read as promoting *the* Good as a ruling principle of a new ideocracy (despite her explicitly arguing that the kind of unity displayed by morality is "of a peculiar kind and quite unlike the closed theoretical unity of the ideologies" [1999, 377]). After all, this is, as we have seen, precisely how Plato has tended to be interpreted and adopted by the canon of Western philosophy and also lies at the basis of claims that his philosophy is totalitarian, that he is, in Popper's (1969) terms, an enemy of the open society, because he sets absolute ideas above political practices.

Any such claim would be difficult to sustain against Murdoch despite her obvious indebtedness to Plato's philosophy. But the espousal of the sovereignty of Good might be thought indicative that she, in her own way, wants to make ethics politically inviolable, just as Plato did. In Murdoch's case, this is because, while the complex emotional intensities, understandings, and social relations associated with ethics certainly have wider resonance with others' experiences, they are initially singular, phenomenologically experienced, moments of unselfing embedded in individual lives and therefore conceptually and politically irreducible events. (They also give access to a reality, including a natural reality, which is itself never reducible to political definitions.) Such complexities may, of course, find themselves axiomatically simplified and expressed in the public sphere, but even here the purpose of such axioms (for example, in terms of concepts of individual human rights) is, Murdoch thinks, "to render the value of the individual inviolable in the context of public morality" (Antonaccio 2000, 161). In other words, politically adopted moral axioms are supposed to serve ethical purposes in the sense of maintaining a space for singular individuals, as both self-interested and ethical beings. (It is important to remember that politics, for Murdoch, is envisaged largely in terms of the realm of that self-interested Hobbesian individualism she also considers "axiomatic" [Murdoch in Antonaccio 2000, 161].)

Of course, there may be political debate concerning the importance of specific ethical ideals, but it is neither possible nor desirable for Murdoch that politics define or fix what is Good as such. Ethics is an individual's ongoing task and responsibility, albeit a task that always takes place within a social context and is therefore socially influenced. This division of powers may seem somewhat artificial (and the boundary between them is certainly porous) and is also, as Murdoch recognizes, dependent on the specific historical development of Western (and specifically liberal humanist) understandings of ethics, politics, and individuals. Not surprisingly, then, when Murdoch claims that axioms "arise out of and refer to a general conception of *human nature* such as *civilized* societies have gradually generated" (Murdoch in Antonaccio 2000, 161, my emphasis), it is easy to spot the historical residues of the anthropological machine at work with all its attendant dangers.

The question remains whether such axioms can or should actually be set aside as necessary givens and protected from secularizing political, as well as ethical, critiques. Do they serve to protect the open texture of social life and/or actually foreclose it in the very name of securing such openness? This is itself a complicated ethicopolitical question that goes to the very core of political theory in the sense that it asks whether this purportedly ethical framework for politics is itself to be politically justified or, alternatively, has ultimate authority (sovereignty) over politics. This question has often been avoided rather than answered by positing naturalistic myths of a state of nature and/or historicist myths of social progress, both of which deploy the anthropological machine in order to place particular "settlements" (in Latour's sense) beyond all critiques—whether ethical or political (see later in this chapter). Murdoch merely notes that "liberal political thinking cannot dispense with the inflexibility of axiomatic morality" and that "we 'cut off the road to an explanation' in order to safeguard the purity of the value, and remove it from vulnerability to certain kinds of argument" (Murdoch 1992, 386).

But, in placing certain moral axioms beyond political debate, what for Murdoch is today a tactical, pragmatic, and historically influenced decision concerning their political inviolability comes uncomfortably close to mimicking the rationales provided for more restrictive models of the pastorate. She is, despite her understanding of ethics, too close to the antipolitical stance that Arendt thinks Plato's metaphysics exemplify. Placing moral axioms beyond and above politics (arguing for the sovereignty of Good) all too easily slips into the kind of moral absolutism that enjoins the populace to obey the rules set out by those who define themselves as having superior knowledge of the True and the Good—a model that subjects most inhabitants to a paternalistic model of household management, that is, of stewardship under *The Laws*. In Arendt's (2005, 12) words "Plato designed his tyranny of truth, in which it is not what is temporally good, of which men can be persuaded, but eternal truth, of which men cannot be persuaded, that is to rule the city." Obviously, Plato's recourse to a form of philosophical absolutism was related to his antidemocratic belief that most people are incapable of coming to their own understanding of ethics. This is far from Murdoch's position. But Arendt's point here is that, on discovering that philosophical arguments and persuasion did not necessarily carry the day, even where Athens's citizens (that is, those whose discussions were constitutive of the political realm) were concerned, Plato's response was to make ethics and philosophy inviolable by extending the model of stewardship beyond the sphere of the household to hold sway over politics itself.

Arendt suggests that Plato argued for the political authority of idea(l)s and their philosophical interpreters as a reaction against the condemnation of Socrates. Dismayed by the outcome of Socrates's trial, Plato sought to give philosophy an authoritative status capable of withstanding the power exercised by mere opinion *(doxa)* among the citizenry, thereby protecting philosophers against those who might treat them as a "common laughing-stock—as Thales was laughed at by a peasant girl" (Arendt 2005, 9). For Plato, the dignity of dialectically revealed truth is besmirched if it becomes subject to mere politicking.

"It is as though the moment the eternal is brought into the midst of men it becomes temporal" (Arendt 2005, 12), open to discussion, just one more opinion among others, and this is what no one who wishes to claim sovereignty for their own ideas can abide. This is, however, as Arendt points out, precisely what a genuinely democratic politics requires.

Despite his best efforts during his trial, Socrates failed to persuade his fellow citizens that the polis needed people like him, and this came to be taken by Plato as a fault with Socrates's understanding of philosophy as an art of political persuasion, as a form of speech lacking compulsion.[19] Plato's *Phaedo* can therefore, Arendt (2005, 7) argues, be read as a "'revised apology,' which he called, with irony, 'more persuasive' *(pithanoteron),* since it ends with the myth of the Hereafter, complete with bodily punishments and rewards, calculated to frighten rather than merely persuade the audience." The real irony here is that, from this perspective, "Plato himself was the first to use the ideas for political purposes, that is, to introduce absolute standards into the realm of human affairs" (8). In doing this, he actually betrays Socrates's legacy, because Socrates knew that part of the price for recognizing plurality and relying on persuasion is that persuasion will inevitably sometimes fail and results in far from perfect agreements. But this imperfection, this recognition of the importance of persuasion without compulsion, was precisely what Socrates was willing to die for when, by the narrowest of margins, Athenians decided that they did not, after all, need his troubling interventions and chose to place their own self-interests above politics as such.

Plato's solution to the tensions between ethics, politics, and philosophy leads to the development of a philosophical form of sovereign (political) power, as a mode of stewardship exercised in the name of the Good. This may be intended to protect the practice of philosophy and the idea(l) of the Good from becoming subject to arbitrary political decision, but it does much more than this. It makes politics, which Arendt regards as the realm of diversity and freedom, subject to what, for all but the elite wielding power, are politically unquestionable practices and decisions and turns ethics into abstract metaphysical ideals in the strong (nonmetaphorical) sense of the term. Ethics is reduced to a form of moral compulsion rather than comprising the very fabric of our lived worldly relations with others. This moral ideal is imposed in the form of relatively fixed norms, rules, and laws, that the "steward" deems accord with the Good. But since Good (ethics as such) is not actually some thing or some absolute value that one can simply be in accordance with, since it is a relation of infinity, not totality, a continual holding open of questioning, this inevitably also alters the very form and understanding of ethics.

The danger in this strategy is one of reifying the concept of the Good, making this idea into both an ahistorical abstract principle and a concretely defined reality that then comes to exercise its own external (moral) authority over life. In Lukács's (1983, 160) terms, it is *philosophically immortalized*. It makes compulsion, or at least the possibility of compulsion, where those wielding power (the philosopher-kings and their later instantiations) decide it is necessary or justified, the keystone of a political, ethical, and philosophical order. It threatens to reduce the freedoms associated with ethics as such to compliance with moral norms, to following rules, and (as Foucault associates with all forms of the pastorate) to an internalized relation of dependence that confines moral feeling to a self-monitoring compliance of each individual with those ideals espoused by the ruling powers. (In other words, it regards the [in Arendt's terms] "a-political" requirement for obedience to the steward's directions as a political virtue.) This is precisely the kind of theoretical and actual tyranny that Murdoch recognizes as a threat to the open texture of ethics as the "abode of freedom" and to politics itself as Arendt (2005, 108) understands it, since for her the "meaning of politics is freedom."

This (metaphysical) reification of the Good as an abstract (totalizing) moral principle lies at the very center of claims to political sovereignty, understood as the ability to suspend the political on the basis of a higher calling and knowledge of what *ought* to be done, that is, a claim to possess access to politically inaccessible truth(s). The "ought" of moral compliance displaces ethics as such and becomes an instrument employed to rule over the political actions and activities of all beings under its auspices. From Arendt's perspective, Plato's attempt to make philosophy politically inviolable actually inaugurates an antipolitical philosophical tradition that is associated with the origins and subsequent possibility of political and moral totalitarianism.[20] This antipolitics is based in a "conceptual framework hostile to popular participation, human diversity (what Arendt dubs "plurality"), and the open-ended debate between equals [a framework that traced back to Plato] came to provide the basic conceptual architecture of Western political thought" (Villa 2000, 7). Here too the apolitical image of stewardship and the household predominates. "Already in Plato the implications for action of this image of the household are clearly indicated: 'For the truly kingly science [of statesmanship] ought not itself to act *[prattein]* but rule *[archein]* over those who can and do act.' It causes them to act, 'for it perceives the beginning and principle *[archē]* of what is necessary for the polis, while the others do only what they are told to do'" (Arendt 2005, 91, quoting Plato's *Statesman,* 305d). *Archē,* meaning both "beginning" and "rule," subsequently comes to be interpreted in terms of "ruling principles" that hold supervisory sway over politics as such.

If the moralistic enclosure of the space of political action depends on a distorting reification of ethics, then we should also recall that ethics *as such,* in Murdoch's and Levinas's sense, might actually be thought of as *anarchic* (which is not, by any means, to suggest that Murdoch or Levinas are *political* anarchists). Their understandings of ethics are anarchic in the sense that ethics is initiated through individuals' concerned involvement with the "singular" reality of others' being and that it marks a refusal to subsume such relations under any absolute ruling principles, rules, or concepts.[21] This anarchic understanding of ethics as an open texture of responsible engagement certainly influences, and is influenced by, similar understandings of politics emphasizing both individual responsibility and openness like Arendt's own (though again these are not necessarily associated with political anarchism as such). In this sense, Murdoch, whose writings focus almost entirely on ethics rather than politics, often expresses views remarkably similar to Arendt. For example, she speaks of the "danger represented by what is called the 'managerial society'" (1999, 180). She also argues that "in the context of political argument and activity the absence of metaphysical background is the point. The successful use of persuasion depends upon a certain waiving of dogma. A good (decent) state, full of active citizens with a vast variety of views and interests, must *preserve* a central area of discussion and reflection wherein differences and individuality are taken for granted. . . . Here there are no authoritarian final arbiters, certainly not God, Reason, or History" (1992, 366). This political understanding, says Murdoch, is both vitally important and "fragile." (There is little, if anything, that Arendt would disagree with here.)

What is clear is that, despite his influence on her work, Murdoch's political intentions are very different from Plato's. Where Plato appears to try to make ethics inviolable by taking it altogether outside and placing it in a position of sovereign power over the political realm, Murdoch actually seeks to retain the open texture of both ethics and politics as parallel, connected, but ultimately irreducible spaces. The former is, roughly speaking, the realm of one's unselfing in relation to the realities of the world, while the latter is a more agonistic realm of discursive self-expression. This means that for her, as for Hume, "good political philosophy is not necessarily good moral philosophy" (Murdoch 1999, 366). Arendt too regards ethics as a matter that "concerns the individual in his singularity" (Arendt quoted in Young-Breuhl 2006, 201) and as being corrupted when made into political principles (205). For Plato, the political authority of the statesman is marked by "his" *ordering* society according to his knowledge of the Good—it is far from democratic. For Murdoch (as with Arendt), the modern state's authority exists only insofar as its sole political purpose is to preserve the space of plurality, difference, and persuasion without compulsion, the open texture of politics and ethics, both of which are ultimately dependent on individuals' taking responsibility for their words and actions.[22] The sovereignty of Good is not, for Murdoch, a principle of political sovereignty; it is an ethical calling.

Situating Levinas

Perhaps we can ecologically recuperate these discussions concerning ethics and politics by considering the extent to which Levinas's understandings might, and might not, inform and be informed by these "anarchic" critiques of political sovereignty and of the metaphysics of the anthropological machine.

For Levinas, the case is slightly different. He rejects the idea that ethics can be captured in terms of any overarching principles and concepts. Ethics is not envisaged, or practiced, primarily as a matter of rational analysis, moral laws, philosophical system building, convergence with tradition, or hedonistic calculus. As with Murdoch, ethics as such is the elicitation of a force pulling us away from our self-centered orbits toward responsibility to others. Paradoxically, then, this least self-centered understanding of ethics becomes deeply personal (in terms of one's relations to Good and to Others), a matter of absolute responsibility for Others.

Ethics are also anarchic in the sense, noted by Peperzak, that for Levinas, ethical responsibility is deemed to be something that arises without and before any definable point of origin; it has no *archē* (beginning) in the ontology of the world. But this placing of ethics "beyond Being," that is, its being regarded as trans- or pre-ontological already assumes a very specific understanding of ontology as a philosophical discourse that concerns itself with producing totalizing accounts of *what there is* in order to allow the selfish and "panoramic [human] *cogito*" (Peperzak 1997, 10) to rule and dominate the world. Such totalizing discourses, Levinas argues, necessarily exclude the very possibility of infinity, of nonclosure, of ethics as such. Ontology, for Levinas, is itself "a manifestation of the natural egoism which constitutes the elementary level of human life" (Peperzak 1997, 10).

But this is itself a rather totalizing view of Western philosophy, which as the case of Bataille exemplifies, has sometimes concerned itself with that which (necessarily) escapes all such totalizing systems. Nor, arguably, is it even entirely fair to Heidegger, whose ontology is Levinas's key target (but see Peperzak 2005, 210–12), given Heidegger's forthright critique of the consequences of the total technological enframing of the world as "standing reserve" or resource (see later in this chapter). It also seems to leave Levinas with a very unworldly (strongly metaphysical) notion of infinity that has more in common with (and, according to Moyn [2005], was heavily influenced by) the negative theology underlying the work of his contemporaries like Karl Barth: contrary to attempts to describe God's attributes, negative theology emphasizes the impossibility of a finite humanity being able to rationally define or comprehend the transcendent alterity, the otherness, of an infinite God.[23] Of course, the transcendent alterity spoken of by Levinas is that of the human face, not of God, but the religious influences on Levinas's work remain pivotal. Not only is ethical responsibility referred to in terms like "height" *(hauteur)* and "transcendence" that retain a strongly metaphysical aura, but even Levinas's own later attempts to secularize his understanding of ethics were, Moyn (2005, 83) argues, ultimately unsuccessful, since the infinity looked for in other people remains, in theological terms, "the divine-in-man."[24]

In other words, as already suggested, Levinas's philosophy is much more indebted to the metaphysics of the anthropological machine than is Murdoch's, and to that extent, less suitable for an ecological ethics precisely because of this failure to secularize the metaphysical (infinite) difference he posits between humans and all other worldly beings. And here another issue arises: If Levinas's ethics is anarchic in the sense that he holds it is impossible to define (ontologically) where ethics begins, then should it not also recognize the comparable difficulty in defining where a specifically human mode of being arises? In limiting the infinity of ethics to intraspecific human relations, Levinas simply takes this anthropological boundary as given. And, as Alphonso Lingis (2005, 107) remarks, why shouldn't "other animals, plants, landscapes, rivers, and clouds, in what is achieved in them, and in their birth (*nature* etymologically is what is born)—in their bare materiality—appeal to us and put demands on us, address their imperatives to us?" Why not recognize, as Murdoch (1999, 381) suggests, that the "indefinability of Good is connected with the unsystematic and inexhaustible variety of the world and the pointlessness of virtue"?

The anthropocentric metaphysics of the Judeo-Christian tradition informing Levinas's work clearly militates against Lingis's and Murdoch's suggestions, as does Levinas's reaction against the political results of Heidegger's secularizing (and, for him, all too worldly) ontological critique of metaphysics. Levinas thought Heidegger's infamous espousal of totalitarian politics in the form of National Socialism was, at least in part, a consequence of the faults in his ontological (and in Levinas's view, totalizing) philosophy. Perhaps Levinas's attempt to place ethics beyond worldly Being might be regarded as a strange reflection and inversion of Plato's previous attempt to make ethics inviolable. If Plato's ideocratic simplification of ethics was a response to the failure of political persuasion (politics as such) that led to his mentor Socrates's condemnation, then perhaps Levinas's ethics was, in part, a response to the serious political failings of his own mentor, Heidegger. Of course, Heidegger's failing was, unlike Socrates's, his own; it was he who proved all too susceptible to the influence of those tyrannical forces seeking to suppress the possibility of politics as such. But this may help in understanding why Levinas's solution is also inverted: rather than making a simplified metaphysical ideal of ethics sovereign over politics in the form of a moralistic (if supposedly paternalistic) tyranny (stewardship under the moral law), Levinas's ethical anarchism resists totalizing/totalitarian tendencies by stressing the indefinable complexity of a relation that calls *every* individual to take responsibility for the (infinite) openness of all (human) others.

However, while anarchic in more than one sense, Levinas, somewhat ironically, retains an unnecessary quasi-theological aspect of compulsion. For Levinas, ethics takes the form of an individualized metaphysical *necessity* concerning our infinite responsibility for others; we are "obsessed and taken hostage for the other" (Peperzak 2005, 36) by a responsibility that comes even before our own being, by the "necessity that the Good choose me first before I can be in a position to choose" (Levinas 2004, 122). Levinas claims that this obsession too is anarchic, because it displaces the idea that ethical subject originates in the "auto-affection of the sovereign ego that would be, after the event, 'compassionate' for another. Quite the contrary: the uniqueness of the responsible ego is possible only *in* being possessed by another, in the trauma suffered prior to any auto-identification, in an unrepresentable *before*. The one affected by the other is an anarchic trauma, or an inspiration of the one by the other" (2004, 123). In other words, this understanding is anarchic in the sense that the ontological priority and temporal authority of the self-interested sovereign subject of modern liberal political and economic theory is dethroned.

But even if we accept, and perhaps we should, that ethical responsibility originates prior to the selfish ego's emergence, why situate this originary matrix of other-directed concern beyond ontology, if by ontology we mean Being in all its infinite and ultimately unfathomable richness? And why speak of this matrix in terms of compulsion rather than as an opening of ethical *possibilities?* This only makes sense if, contra the movement of secularization, one wants to both retain a form of supernatural (but actually anthropocentric) religiosity and, however subtly and circuitously, something of the authority of a moral "ought" that binds us to an inescapable duty to others. As Antonaccio (2000, 222n52) notes, "Murdoch would reject the language of command, lordship, and accusation that pervades Levinas's account of the other's claim on the self, and she would likewise resist his constriction of the domain of ethics to the moral 'ought.'" This is not to say that Murdoch is unconcerned about issues of obligation or duty but that she rightly retains a form of ethical openness in the face of our encounters with others that is more true to life: for her, *Metaphysics* (is only) *a Guide to Morals,* not its be all and end all. "A moral philosophy should be inhabited" (Murdoch 1970, 47).

An anarchic ecological ethics would not be dependent on defining, in any absolute or authoritative form, its worldly origins or tasks, but neither could it seek to escape from such worldliness into a metaphysics that claims to be anything other than a necessarily fallible and questionable guide for finding ways to think about our responsibilities for others, human and more-than-human. Such metaphysics are inevitably conceptual simplifications (since that is their purpose), ultimately emerging through our phenomenal experiences of the world but also calling on particular traditions, knowledge, and histories. They help us feel our way around in the world; they influence how we affect and are affected by it. But to do this and to remain ethical, they must also be open ended, tentative, and subject to revision. When they lay claim to represent the reality of the world or the essence of other beings, then, as Arendt, Levinas, and Murdoch all agree, they threaten to become tyrannical and antithetical to ethics as such. Only a concerned attentiveness to the diversity, plurality, and complexity of the world, and only that *amor mundi,* love for the world, which Arendt remarked on late in her life, can effectively resist this constant temptation.[25] This love cannot be a matter of compulsion (nor like any love is it simply a matter of decision or choice and certainly not the application of abstract principles that managerial forms of stewardship require); it is a response(ability)—a possibility given by and through one's concerned involvement in the world.

An ecological politics, too, has to retain an open texture, to be anarchic in its renunciation of any claim to represent the moral authority or sovereign powers of nature. (As was long supposed to be the case with sharks, politics as such retains its vitality only so long as it keeps swimming.) But this means that while politics and ethics spill over into each other, they are also irreducible, and their movements are not necessarily attuned to each other. One cannot be awarded precedence over the other nor be placed entirely beyond its influence, even though, at their most cooperative, both are modes of expression of individual freedom understood as vital but never entirely attainable ideals of a good life in concert with others. These ideals are utopian not because they exist in some pure form in a metaphysical beyond but precisely because they can offer some guidance to our worldly existence, which is always that of being-in-the-world, never a purified being entirely separable from the world. This is what mortality means and what community involves. We are constitutively impure and incarnate (embodied and worldly beings), and secularization is a movement of thought and practice that recognizes this kenotic reality. Instead of looking for the divine in Man (the metaphysics of the anthropological machine), we might instead try to divine, sense something of (as a water diviner does), the flows and depths of diverse worldly existences happening beneath their surface appearances.

How then could the relation between ethics and politics be re-conceptualized? Perhaps in terms of yet another thesis: *if politics as such is,* in Arendt and Murdoch's sense an open texture of interactions, or in Agamben's (2000) words, *a matter of pure means, a "means without end," then ethics could be thought in terms of being concerned with Others as impure ends, as beings of indefinable (infinite) value but finite worldly existence.*[26]

3 PRIMITIVISM

Anarchy, Politics, and the State of Nature

> All significant concepts of the modern theory of the state are
> secularized theological concepts not only because of their histori-
> cal development—in which they were transferred from theology
> to the theory of the state, whereby, for example, the omnipotent
> God became the omnipotent lawgiver—but also because of their
> systematic structure.
>
> —Carl Schmitt, *Political Theology:*
> *Four Chapters on the Concept of Sovereignty*

The State of Nature

What political possibilities arise in articulating anarchic ethics with
ecology? An obvious question seems to be whether some of these
ethical and ecological possibilities have an affiliation with anarchy in
a political sense. Might anarchists' advocacy of an unfettered open
texture of social relations extend to challenging the idea of political dominion over the more-than-human world? Although many recent anarchists, like Morris (1996, 58), have claimed that "anarchism implies
and incorporates an ecological attitude toward nature," whether and
how it might do so remains a matter of intense debate (Smith 2007b).
After all, many anarchists have traditionally been relatively uncritical
exponents of versions of the anthropological machine (May 1994, 63).
Many more have tended to regard ethics per se with suspicion, as just
another (moralistic) method of ensuring compliance with dominant social norms. On the other hand, anarchy exemplifies (theoretically and
practically) a secularizing critique of all metaphysically posited *origins* and *principles* attempting to justify political authority. And even
though various forms of anarchist politics may be more (Kropotkin) or

less (Stirner) informed by ethical concerns, they all nonetheless ensure that any tendency to use ethical principles in an ideocratic manner or to regard them as matters of compulsion are continually and reflexively critiqued.

If, as the opening quotation from Schmitt suggests, the origins of the state's political omnipotence lies in (partially) secularized theological concepts, then this is also where secularizing anarchist critiques might be invoked. The "state of nature," that theoretical/theological condition posited by early modern political philosophers like Hobbes, Locke, and Rousseau as prior to, originating, and justifying the modern political authority of the state, should then be understood in precisely this way. It is the modern state's successor to the Bible's Edenic myths on which these philosophers implicitly and explicitly drew. And following the pattern initiated by Plato, this myth too is used to justify political authority as an "unfortunate" necessity in a world no longer under a perfect, God-given order. An ecological anarchy, ethically concerned with the current state of nature, the damaged condition of the natural world, might then also need to offer a critique of the mythic state of nature—the ideal and ideological beginnings *(archē)* of the forms of political authority now ruling over this world.

Such a critique could hardly ignore those iterations of the anthropological machine associated with the state of nature, as the mythic origins of state authority coincide with the emergence of a particularly modern concept of the properly or fully human—that is, the abstract idea of a humanity now socially, politically, and historically separated from its natural state. This separation from and dominion over nature is no longer regarded by political philosophy as directly or entirely God given but as a consequence of exercising the capacities inherent in a specifically human nature. The defining moment of those deemed fully human is thereby given a more explicitly political, secular, and quasi-historical twist by being associated with the emergence of civil(ized) society. Despite that, the apolitical essence it posits (human nature), an abstraction supposedly (but impossibly) reiterated and incarnated in every individual human—who is therefore, to this extent, the Same as everyone else—remains metaphysical in the strong sense (a point well made by Althusser [1969, 228–29] in his critique of Marxist humanism). Human nature remains an inadequately secularized principle of absolute identity set over and against real individuals in their plurality and diversity. Like every iteration of the anthropological machine, this abstraction can subsequently be deployed to decide who is and who is not properly human, who rightfully has dominion over the Earth, by those who count themselves civilized.

It is surely ironic that despite their wholesale orientation toward the future, the political institutions of modernity are actually no less dependent than the polis of ancient Greece on myths of origin, metaphysical accounts of a beginning *(archē)* that gives rise to ruling principles— this precontractual state of nature existing prior to modernism's own ideals of a civilized society.[1] What is more, since the purpose of these early modern political theories was to elucidate culturally binding principles of governance and moral law, the state of nature was almost always envisaged as an anarchic and amoral realm. In Locke's words: "To understand Political Power right, and derive it from its Original, we must consider what State all Men are naturally in, and that is, a *State of perfect freedom* to order their Actions and dispose of their Possessions, and Persons as they think fit, within the bounds of the Law of Nature, without asking leave, or depending upon the Will of any other Man" (Locke 1988, 269).

The exact manner in which this prehistoric existence was envisaged depended on the particular theorist's tendency toward an optimistic or pessimistic assessment of human nature and human society. For Hobbes (1960, 82), this anarchic state was famously characterized by "continual fear, and danger of violent death; and the life of man, solitary, poor, nasty, brutish and short." For Rousseau, driven as much by pessimism about the parlous state of his contemporaries as optimism about human nature, this primitive anarchy had distinct advantages. Life was (generally) marked by individual isolation, indolence, robust health, and heart's ease because the "produce of the earth furnished him with all he needed, and instinct told him how to use it" (1986a, 84).[2]

Optimist and pessimist alike agreed that civilization was to be defined in terms of the distinction between nature and culture and by the movement of the latter away from the former. Humanity was driven to distinguish and distance itself from its previously primitive existence, something that could only be achieved through hard work and the employment of that unique human faculty "reason." The irrational anarchism characteristic of the state of nature was superseded, whether from necessity or choice, by a rational agreement, a social contract. This contract was an agreement to enter into the moral and political order of civilization, to limit one's inherent freedoms and control one's inherent nature in the name of reason and social progress.

This then is modernity's key foundational political narrative. It has been employed in numerous ways and to justify diverse political ends, from monarchism to regicide, but its ontological status remains ambiguous. Did this state of nature exist, or was it more imaginary than actual? For Locke (1988, 276), the state of nature was a historical and geographical reality, a matter of established fact: "The world never was, nor ever will be, without Numbers of Men in that State." For Rousseau (1986a, 44), the state of nature "perhaps never did exist, and probably never will exist." It was, more explicitly, a product of modernity's political imaginary: "The philosophers, who have inquired into the foundations of society, have all felt the necessity of going back to a state of nature; but not one of them has got there" (50).

Whatever its ontological status, its ideological effects were extensive and all too easily applied to the real world. And, contra Rousseau, the dominant ideological perspective of modernism continued to regard this divisive yet civilizing movement away from nature in an entirely positive light. This movement comprises, after all, in historicism's terms, precisely what constitutes progress. In this way, the current conditions of those peoples classified as primitive were first used to provide evidence for contrasting speculations. This supposedly progressive anthropological "just-so story" could then be turned back on itself and pressed into service to justify the brutal treatment of those same (not properly human) peoples—John Locke himself had financial interests in the slave trade.

Locke's own version of this story mentions three critical moments, which mark stages in the change from a state of nature to that of civil society. The first is the appropriation of nature, transforming it from God's common gift to humankind to personal property; the second is the invention of money; the third is the social contract itself. These three ideological moments represent a movement from the theological dominion of nature to a world increasingly reenvisaged in terms of circulating capital and state sovereignty. In the first instance, nature is altered through the admixture of human labor. Since the "Labour of his Body, and the *Work* of his Hands we may say, are properly his [individual property]. Whatsoever then he removes out of the State of Nature hath provided, and left it in, he hath mixed his *Labour* with, and joyned to it something that is his own, and thereby makes it his *Property*" (Locke 1988, 288). Nature thus becomes parceled up; wilderness becomes tamed, domesticated, transformed, and owned through individual labor. (Though even Locke seems politically and ecologically astute enough to have added the proviso that this holds true only where there is "enough, and as good left in common for others" [288].) Labor has "put a distinction between" the commonalty of natural objects and personal property, it has "added something to them more than Nature" (288). In instrumental terms, this addition is also a necessary improvement "without which the common is no use" (289). (It is, however, surely ironic that Locke, so familiar from his Puritan upbringing with the Bible's Edenic narratives, should choose to illustrate his case for private property with the example of picking apples.)

The invention of money allows a second qualitative change to take place, because it marks both the beginning of the commodification of nature and the introduction of a hierarchical social organization. Originally, the extent of individuals' personal property was limited by their labor power, by the amount of land it was physically possible for them to make use of, which "did confine every Man's *Possession,* to a very moderate Proportion" (292). But "the *Invention of Money,* and the tacit Agreement of Men to put a value on it, introduced (by Consent) larger Possessions, and a Right to them" (293). Money, unlike nature's products, does not spoil; it can be stored and accumulated indefinitely, and so if people consent to take money "in exchange for the truly useful, but perishable Supports of Life" (301), then by default they have "agreed to disproportionate and unequal Possession of the Earth" (302). From Locke's perspective, civil(ized) society arises out of the need to protect inequalities. There is little point in stealing others' perishable property if one already has all one can use, but money provides an imperishable motive. And so, in response to this new situation, people sign up to the social contract. They agree to give up their natural freedoms and to submit to the authority of "a common establish'd Law and Judicature . . . with Authority to decide Controversies between them, and punish Offenders" (324). This contract is, Locke makes plain, the final and most important aspect demarcating civil society from the state of nature.

There is, though, another, less explicit element that Locke adds to these three originating moments: the transformation of nature by the admixture of human labor, the commodification of nature through its symbolic incorporation in a monetary economy, and the development of a hierarchical system of rational political/legal authority. This additional element might be described as a "work ethic." Since it is human labor that is presumed to improve (transform and add monetary value to) nature, then productive labor becomes a moral duty of civilization's citizens, idleness and unemployment a sin. Peoples' failure to rise above their supposedly primitive state also implies a moral failing on their own part.

These mutually supportive and interacting elements constitute the conditions for that overarching ideology of progress that comes to pervade every aspect of modern life.[3] And so long as modernity continues to deliver material benefits, this idea remains more or less immune to political critique. It forms part of the historicist background (the second nature) of modernist culture. As Marcuse (1991, 1) argues, an apolitical, "comfortable, smooth, reasonable, democratic unfreedom predominates in industrial civilisation, a token of technical progress." It is only when things go wrong, fail to meet expectations, or run into unexpected opposition, that "the social world loses its character as a natural phenomenon that the question of its natural or conventional character . . . of social facts can be posed" (Bourdieu 1991, 169). Today, though, it is precisely the proliferation of ecological risks, a direct consequence of an unquestioning adherence to the rationality of technological progress, that now provokes political challenges to modernism's own foundational myths. These risks expose the political character of our taken-for-granted ideas concerning social conventions and nature itself.

Primitivism

The emergence of an ecologically informed politics of "anarcho-primitivism" that offers a fundamental critique of modern state authority and of human dominion over nature might be understood in this light. This explicitly primitivist strand in environmental anarchism, which coalesced around journals like *Fifth Estate*, (the UK and North American editions of) *Green Anarchist*, and *Anarchy: A Journal of Desire Armed* and the writings of, among others, Fredy Perlman (1983) and John Zerzan (1994; 1999), has gone largely unnoticed by political philosophers. Yet its appearance should not be entirely unexpected because, as Passmore (1974, 38; see chapter 1 of the book in hand) suggested, the opposite pole of ecological dominion has always been primitivism in one form or another.

While those associated with primitivism hold a variety of perspectives and sometimes question or even eschew the label "primitivism,"[4] their arguments share a family resemblance in terms of their critiques of civilization and technology, which are regarded as instigating and perpetuating both social inequalities and environmental crises. Primitivism is unusual in tracing these problems right back to modern society's underlying principles and origins *(archē)*. The destructive repercussions of the myth of progress were, Zerzan (1994, 152) argues, present "from history's very beginning. With the emergence of agriculture and civilization commenced, for instance, the progressive destruction of nature." Primitivists regard modern society—and in many cases civilization in general—as inherently, rather than accidentally, destructive. The oil slicks polluting Puget Sound, the clouds of radioactivity released from Chernobyl, devastating mudslides from deforested hillsides, the ozone hole, global warming, asthma-inducing smogs, and so on, are accidental only in the very trivial sense that they were not (usually) the intended consequences of the social activities concerned. These events, though often unforeseen, are by no means accidental by-products of modernity but a necessary and inevitable corollary of modern modes of production, of progress itself.

Insofar as primitivists regard civilization as inherently and irredeemably destructive, the scope of their critique is similarly all encompassing. Where "ideologies such as Marxism, classical anarchism and feminism oppose aspects of civilization, only anarcho-primitivism opposes civilization, the context within which the various forms of oppression proliferate and become pervasive—and, indeed, possible."[5] Primitivists call into question every aspect of a culture where "private property, industrial medicine and food, computer technology, mass media, representative government, etc., all work together to maintain our alienation from wildness" (Black and Green Network, n.d., 1).

Most, though not necessarily all, primitivists (see William 2001, 39) *nostalga* refer positively to a precivilized past presumed to have existed before settled patterns of agriculture emerged. These Paleolithic gatherer-hunter communities, like those who created the Lascaux paintings (see chapter 1), are supposed to have experienced lives of "primitive affluence," much as envisaged by Marshall Sahlins in his influential *Stone Age Economics* (1972). "Life before domestication/agriculture" says Zerzan (1994, 16), "was in fact largely one of leisure, intimacy with nature, sensual wisdom, sexual equality, and health. This was our human nature, for a couple of million years, prior to our enslavement by priests, kings, and bosses." It was not lack of intelligence or lack of ambition that stopped Paleolithic cultures from "advancing"; rather, "the success and satisfaction of a gatherer-hunter existence is the very reason for the pronounced absence of 'progress'" (23). Such societies were, we are told, nonhierarchical, largely nonviolent and noncompetitive, had no conception of private property and had inordinate amounts of free time, which they spent socializing. Paleolithic peoples were healthier and happier in complete contrast to the current "landscape of absence . . . the hollow cycle of consumerism and the mediated emptiness of high-tech dependency" (144).

Primitivism completely inverts the humanist's progressive interpretation of world history. The state of nature is reinterpreted as a prehistoric condition of relative, if not absolute, ecological and anarchic social harmony, *almost* an anarchic Eden.[6] Movement away from these gatherer-hunter societies, which were embedded in wild nature, constitutes the anarcho-primitivist equivalent of the biblical Fall. The key moments in Locke's political imagery are recuperated and reevaluated entirely negatively as exemplifying the increasingly pervasive contamination of purer, original relations with a still-perfect natural world. Civilization is marked by *ecological contamination*—the pollution and destruction of wilderness and its transformation into a resource for human-labor *economic contamination* marked by the commodification of the life-world and the massive and immoderate increase in the consumption of the natural world this allows and promotes—and by *ethicopolitical contamination* of previously unrestricted freedoms by an ideology and discourses of moral and political governance justified through the myth of the social contract.

Primitivism attempts to recuperate the purity of the state of nature by rejecting this "culture of contamination" in its entirety. It turns the idea of progress on its head in every sense—technological, social, political, and ethical. In opposition to civilization's search for rational order in all these spheres, it emphasizes the play of "instinct"—direct, nonlinguistically mediated personal experiences—and "intuition as a crucial part of rewilding" individuals (Ardilla 2003, 46). Moral authority is replaced by an individually liberating and ecologically sustainable reimmersion in wild(er)ness. The alternatives are, Zerzan (1994, 146) argues, to carry on toward increasing domestication and alienation or "turn in the direction of joyful upheaval, passionate and feral embrace of wildness and life." In Derrick Jensen's (2004, 18) words: "I want to live in a world with more wild salmon every year than the year before, more migratory songbirds, more natural forest communities, more fish in the ocean, less dioxin in every mother's breast milk. And I'll do what it takes to get there. And what it will take is for us to dismantle every-thing we see around us. It will take, at the very least, the destruction of civilization, which has been killing the planet for 6000 years. If that's primitivism then I guess I'm a primitivist."

The primitivist aim is to free people from "the historical perspective that holds us captive and fall again into the cyclical patterns that char-acterize the natural world" (Jensen 2004, 22). For Jensen, this means, quite literally, an eventual return to a stone-age existence. Zerzan (1994, 46) too contrasts an idyllic vision of "authentic," unmediated past relations to nature with the social and ecological emptiness of the present and the possibilities of finding again what once was in a *Future Primitive*. Such explicitly immoderate ideals make primitivism an easy political target, and critics have declared this largely uncritical celebration of prehistoric (and/or supposedly primitive) societies and primitivism's blanket condemnation of every aspect of modern civili-zation irrational and regressive. As one primitivist notes:

> When we say we want green anarchy, a stateless society, free and in har-mony with Nature, people tell us that it's a nice dream but it'll never happen as "it's against human nature." The point is that it *has* happened—green anarchy was how *all* people lived for a good 90% of history . . . how some *still* live better than we do today. When we point this out, people start pissing and whining about "going back to the caves" and getting protective about their TVs, cars and other fruits of "Progress," particularly Lefties and "anarchists" who don't know the difference and who think "Progress" is some inevitable law of Nature and not part and parcel of State society and the self-serving elites ruling it. We'll demolish those myths.[7]

Of course, it might be asked what the point of demolishing one set of myths is if they are simply replaced with another? What, if anything, is critical or secularizing about such a strategy? But it is also interesting that the most vehement criticisms of primitivism have come from anarchists who want to defend the metaphysical and historicist presuppositions it challenges, those who associate themselves with Enlightenment notions of human nature, rationality, and even, in Murray Bookchin's case, with a quasi-Hegelian (historicist) teleology of social progress.

Bookchin (1995a) was a key and early proponent of radical ecology. Just before Rachel Carson's *Silent Spring* hit the bookstores in 1962, Bookchin (under the pseudonym Lewis Herber) was publishing his own critique of *Our Synthetic Environment*, exposing the reckless use of pesticides like DDT, the dangers of feeding hormones to domestic livestock, the health effects of excessive urbanization, and the dangers of low-level radioactivity. In a bizarrely prescient 1966 article in the journal *Anarchy*, entitled "Ecology and Revolutionary Thought" (the first version of which appeared in 1964), he even suggests that the "mounting blanket of carbon dioxide, by intercepting heat radiated from the earth into outer space, leads to rising atmospheric temperatures, to a more violent circulation of air, to more destructive storm patterns, and eventually . . . to a melting of the polar ice caps" (Herber 1966, 323; see also Herber 1962).

However, in his later years, Bookchin attacked what he disparagingly labeled the "lifestyle anarchism" of writers like Zerzan as symptomatic of a contemporary situation "plagued by the advent of . . . an anti-Enlightenment culture with psychologistic, mystical, antirational, and quasi-religious overtones . . . [where] the ecology movement risks the prospect of becoming a haven for primitivism and nature mysticism" (Bookchin 1999).[8] He reiterated these worries in his critique of the work of David Watson (aka George Bradford), summarizing Watson's stance as one in which "redemption can be achieved only by regression. The rise of civilization becomes humanity's great lapse, its Fall from Eden, and 'our humanity' can be 'reclaimed' only through a prelapsarian return to the lost Eden, through recovery rather than discovery" (Bookchin 1999). Watson and the *Fifth Estate* group with which he is associated exemplify, says Bookchin (1995a, 26–27) a "mystical and irrationalist anarchism." "What is arresting in [*Fifth Estate*'s] periodical is the primitivistic, prerational, antitechnological, and anticivilizational cult that lies at the core of its articles."[9]

Ironically, though, this idealization of the distant past is not unique to primitivism. Bookchin himself often alluded to the existence of more egalitarian prehistoric societies characterized by avoidance of coercion and "complete parity" among individuals, age groups, and sexes, together with "their high respect for the natural world and the members of their communities" (1982, 56–58). It remains unclear how or why he distinguished his utopian view of Neolithic farming communities from the "simplistic" and "regressive" views of those who regard Paleolithic gatherer-hunter societies as similarly egalitarian. Indeed, White (2003) traces the way in which in later editions of *The Ecology of Freedom*, and especially in his publications from the mid-1990s onward, Bookchin increasingly distanced himself from such speculations. "Appalled by the growth of avowed 'primitivist' and even 'anti-civilizationalist' currents in American anarchist circles, Bookchin . . . appeared simply concerned to refute those who would seek to 'substitute mythic notions of a pristine and primitive past that probably never existed'" (White 2003, 46, quoting Bookchin 1995b, 122). It seems that Bookchin's position, for all his polemical excesses, might once have shared more with primitivism than he later admits (Best 1998).

There are good reasons to think that anarchists, no less than proponents of other political positions, have frequently looked to accounts of a primitive state of nature (or what they wrongly considered to be its contemporary equivalents) to justify and exemplify their politics. Woodcock (1975, 22), for example, refers to that "antique vision" that draws anarchists "nostalgically to a contemplation of man as he may have been in those fragments of a libertarian past . . . an attitude which not only seeks to establish a continuity—almost a tradition—uniting all non-authoritarian societies, but also regards simplicity of life and nearness to nature as positive virtues." The concern here is that a misplaced nostalgia mistakes such Edenic narratives as accurate descriptions of prehistoric reality rather than recognizing their present role as exercising a critical (secularizing) political imaginary. That which begins as a critique of metaphysical absolutes (especially Hobbesian notions of human nature) and supposedly archaic principles of private property, commodification and political (state) authority ends by claiming (pre)historical authority for its own version of events. What starts as political critique begins to take on the guise of a political *program*— one that comes perilously close to being just another (anti-Hegelian) version of an end of history thesis—"Some day history will come to an end" (Jensen 2004, 22). This naïve "realism," this failure to appreciate the ambiguous role of the political imaginary, is why anarcho-primitivism is readily dismissed as an extreme, crazy, and impractical form of radical environmentalism and why the political constellation of possibilities it opens are sometimes hidden under squabbles about the relative merits of an (impossible) return to Paleolithic lifestyles.

The state of nature is not something we can or did actually inhabit, not a prehistoric reality or a future possibility. Ironically, like Locke, and contra Rousseau, some primitivists suggest that such a state is a straightforwardly mundane reality rather than a meaningful mythic account of the human predicament. They call on archaeological evidence and employ anthropological accounts of contemporary indigenous populations to support the everyday reality of past and present primitivist communities.[10] But, as we have already seen, Eliade (1987, 95) argues this is not the purpose of a myth: to "tell a myth is to proclaim what happened *ab origine*" and thereby make of this telling an "apodictic truth," a "sacred reality." The myth of leaving behind of a state of nature through progress is the apodictic truth, the sacred modernist shibboleth that (paradoxically) underlies its program of theological desacrilization. Reading this myth or its antithesis literally (as Christian fundamentalists read theological myths of origin) means that the primitivists find themselves limited to reversing modernity's evaluations, thus explicitly endorsing the modernist dichotomy between (the state of) nature and culture. Insofar as primitivists emphasize the reality of this return of the primitive, they risk prehistory repeating itself in a manner both tragic and farcical.

Yet, insofar as primitivism, like all forms of anarchy, also employs a "hermeneutics of suspicion" that strives to be attentive to the dangers of *all* political programs and the resurrection of *all* metaphysical absolutes, this kind of tragic–comic repetition is avoidable. Primitivism's reversal of Lockean and early modern philosophical myths is politically important and not just because it signifies contemporary disaffection with modern sociopolitical formations (especially capitalism and the nation-state). In tackling the productivist and contractarian myth of political authority's origins head-on, primitivism provides a countermodern critique of the very discourses that served to justify human dominion over the natural world. Thus, despite appearances, anarcho-primitivism is not a political side issue—something to be thoughtlessly disregarded—but a radically heterodox politics intent on challenging modernism's most settled opinions *(doxa)*, especially the supposedly unquestionable assumption of human dominion. It goes to the (metaphysical) roots of the secularized theological concepts underlying state authority, commodification, and the (moralistic) ideology of the human proprietorship of nature achieved though labor.

In this sense, and like various forms of radical ecology, anarcho-primitivism does not artificially limit its political critique to the boundaries defined by the anthropological machine. It suggests that we should not just settle, for example, for those accounts of the alienating effects of commodity fetishism associated with the predominance of exchange values, accounts that define the anthropological limit of many Marxist takes on our environmental problems (for example, Biro 2005). Anarcho-primitivism rejects not only the commodification of nature but also the very idea of a specifically human form of labor that automatically stamps nature with a seal of proprietorship. It encourages us to ask why the admixture of human labor should, after all, be regarded as justifying a hierarchic, one-directional *expropriation* of any aspect of nature deemed necessary and as simultaneously comprising the sole active element in eliciting nature's value. This reduction of nature to its actual or potential use values to humans may provide a (dubious) ground for criticizing the disembodied circulation of capital in the form of abstract exchange values—a circulation that undoubtedly has extremely corrosive environmental effects (Kovel 2002)—but it also redefines nature as nothing more than a human resource, as a body of material subjected to human dominion/stewardship (Smith 2001a, chapter 3). In Marx's own words, "Labour is, first of all, a *process* between man and nature, a process by which man, through his own actions, mediates, *regulates* and *controls* the metabolism of nature as a force of nature. . . . He develops the potentialities slumbering within nature and subjects the play of its forces to his own *sovereign* power" (1990, 284, my emphasis).

But why is human labor, always and alone, presumed to be so decisive? This absolute distinction between the results of human labor and the works of nature seems (following and extending Schmitt's suggestion) consequent on accepting insufficiently secularized theological concepts: first, the *mythic* supposition that the world has, from the beginning *(ab origine),* been given in its entirety to humanity—whether or not God is explicitly invoked, as in Locke's case, and second, through the more or less surreptitious deployment of metaphysical idea(l)s of human nature. These idea(l)s define specifically human capacities that are always already presumed to have a higher, elevated, indeed quasi-mystical quality about them. Spectral shreds of theological thinking thus cling to even the most materialistic conceptualizations of humans' mundane practices when and where these become reified as first principles *(archē)*. And so, where human labor is called upon to play a defining (anthropological) role, it too comes to be spoken of in awed tones as inspiring, giving life to, and producing definite form to a purportedly slumbering, dead, dark, and/or inchoate nature, much as the word *(logos)* of god in Genesis does, that is, at the anthropocentrically imagined beginning (origin) of all creation.

Thus, even Marx, who, shifting away from his early concept of "species-being," tellingly criticized Feuerbach's metaphysical notion of human nature, still grants human (social) labor special powers.[11] In the *Grundrisse,* he describes how *"living* labour, through its *realization in the material, transforms* it, and this transformation . . . maintains the material in a *definite* form, and *subjects* changes in that form to the *purpose* of labour. Labour is the *living, shaping, fire"* (Marx in Schmidt 1971, 76, my emphasis). This admixture of human labor is mysterious and controlling in its animating effects but quite contrary to the sympathetic magics practiced at Lascaux, for such modern anthropological machinations no longer imply any ethical responsibility at all for the more-than-human world. Rather, they take it for granted that nature as a whole is made present for humanity to use as it sees fit.[12] As Alfred Schmidt (1971) suggests, human labor transforms the "in-itself" of nature into a "for-us" (see Smith 2001a, chapter 3). Here, too, labor takes on a moral overtone as praiseworthy and socially elevating. Marx, as Baudrillard (1992, 107) suggests, seems guilty of an "aberrant sanctification of work,"[13] a sanctification evidenced in the work "ethic" that underlies Marx's own teleological account of the historical movement of humanity away from its "animal nature" through stateless "primitive communism," no less than it is evidenced in its Lockean, liberal, and capitalist equivalents.

This quasi-theological acceptance of the special status of human labor is, at root, as arbitrary, partial, and metaphysical as the claims Levinas makes about the unique responsibilities elicited by the human face (see chapter 2). Both seem to carry within them a residual notion of the divine-in-human and exemplify an anthropological refusal to countenance the possibility of ethical or political concerns for the more-than-human world. By contrast, despite its theoretical naivety and its practical impossibility, primitivism's political inversion of the state of nature at least reveals the mythic origination of the claim that labor accords humanity proprietorship over the natural world. More than this, by refusing to reduce nature to either commodity or resource, primitivism opens certain political, ethical, and ecological possibilities that might become constitutive of a radical ecological politics, possibilities it articulates in terms of anarchism's antiauthoritarian concerns to maintain open and diverse political textures. Of course, this still leaves many serious questions concerning how radical ecology might actually envisage the political articulation of ethics and ecology and how to relate the critique of human dominion over nature to that of modern state authority, that is, to the Leviathan's claims to sovereignty.

Innocence

> Through a curious reversal peculiar to our age, it is innocence that is
> called on to justify itself.
>
> —Albert Camus, *The Rebel*

If, as Plato's Stranger relates to Socrates, the inhabitants of the "Golden Age" living under the stewardship of the gods needed no political constitution (see chapter 1), then the stateless "future primitive" condition envisaged by writers like Zerzan also seems to echo such an apolitical utopian myth. Of course, this end is not envisaged in terms of any kind of rule by gods or (human) masters. There would only be, in Locke's (1988, 269) already quoted words, a "*state* [condition] *of perfect freedom* [for individuals] to order their Actions and dispose of their . . . Persons as they think fit, within the bounds of the Law of Nature." But this condition, when characterized, for example, in terms of Jensen's "fall" into the supposedly cyclical patterns of the natural world, seems far from desirable if, as the passage from Locke suggests, it thereby assumes that people should simply be subject to natural laws. We need to ask, in what sense and to what extent is being subject to the laws (or cycles) of nature actually compatible with any kind of freedom, especially when this freedom is defined in terms of an absence of (freedom from, not freedom through) politics? This is a vital question and one that echoes some of the concerns of Bookchin and other critics of primitivism. It also animates critics of deep and radical ecologists, since these too have often employed similar language concerning the freedoms experienced in unmediated involvements in nature, especially "wilderness."

To be clear, primitivists like Jensen are not in any sense apolitical or against the kind of political collaboration they believe necessary to resist civilization—far from it (Jensen 2006, 890). Rather than regarding politics as a pure means (as an open texture of free association in Agamben or Arendt's sense), they see such political involvements as a means of (an "endgame" for) bringing about a utopian condition, an end of history, where politics as such will disappear. After all, primitivists are not alone in thinking that politics is inextricably associated with civilization, with the city *(polis),* rather than existence within natural communities. To the extent that primitivism remains caught within the same (albeit inverted) mythic framework of the state of nature, its desired ends would almost inevitably suggest a tacit acceptance of the sovereignty of nature's laws over posthistoric human life. If early modern philosophers like Locke use the myth of the state of nature to naturalize a certain political order (Latour's modern settlement; see Introduction chapter 5), to say that this order originates in a natural order beyond political critique, many primitivists apparently deploy the same stateless condition to imply that politics as such is an unnatural imposition. Such a formula would, ironically, retain an absolute anthropological distinction between nature and (political) culture. It would also seem somewhat ridiculous to exhort people to struggle against all forms of human political authority yet to do nothing politically to organize themselves in response to nature's myriad impositions on their lives.

This same disjunction between means and ends is somewhat less apparent when it comes to primitivism's relation to ethics rather than politics because, while the outright rejection of politics would usually be taken as indicative of an apathetic refusal to resist oppression, the rejection of ethics (at least in the form of oppressive and repressive moral norms) *could* be interpreted as a transgressive achievement in itself. Claiming to be against ethics, in the present as well as some future primitive condition, may actually be associated with a certain radical cachet. What matters, then, is whether the critique of repressive aspects of morality is itself regarded as emerging from, and expressing, ethical concerns for others or as part of a wholesale critique of ethics as such now framed only as a constraining influence on individual liberty. Primitivism can be read in both ways. Insofar as primitivists buy into the state of nature as a potentially inhabitable reality, they tend to view this condition as amoral, as exemplifying the overcoming of ethics as such. This end, then, influences their current practices. By contrast, insofar as the state of nature is regarded as a myth with political and ethical intent, there is scope to distinguish between socially oppressive aspects of morality and ethics as such (in the sense of the more secular understanding of Murdoch or Levinas's work discussed in chapter 2). Indeed, from this latter perspective, ethics as such might rightfully be regarded as a critical aspect of, and inspirational source for, any form of life that wishes to assert the value of freedom or of Other individuals.

At issue, then, is the nature of primitivism's all-or-nothing approach. Civilization, understood as a culture of contamination, is held to have sullied a condition of primal (almost Edenic) innocence in a state of nature. As always, innocence is that original condition which is thought of as external to, or preceding entrance into, any moral order. As Kierkegaard (1980, 44) argued, knowledge of good and evil is the sign of Adam's loss of innocence, it is a distinction that can only "follow as a consequence of the enjoyment of the [forbidden] fruit" of knowledge. Innocence is therefore amoral, not immoral, neither knowing nor respecting right or wrong but acting only according to desire and need. "It appears," says Rousseau (1986b, 71), "at first view, that men in a state of nature, having no moral relations or determinate obligations one with another could not be either good or bad, virtuous or vicious." The apolitical isolation of these individuals is compounded by their amorality, and while Rousseau recognizes that such amorality may seem terrible to his "civilized" readers, they must not prejudge such issues. Rather, people should look at their society and consider whether or not "postcontractual" life is any better— "whether virtues or vices preponderate among civilized men: and whether their virtues do them more good than their vices do harm" (Rousseau 1986b, 71).

Although the theological associations of a term like *innocence* (its connotations of spiritual purity and passivity) mean that it rarely appears explicitly in anarcho-primitivist texts, it remains a key subtext pervading accounts of that which has been lost through the corrosive contacts with progress. It is a condition that survives today, if at all, only at the margins of civilization or as a future idea(l) of amorality that can be achieved by shedding the residual feelings of moralizing guilt imposed on human behavior by civilization. This view of morality as socially and psychologically constraining is shared by most

anarchists to some degree, but primitivists tend to go further.[14] In Feral
Faun's (2000, 72, my emphasis) terms, morality is one of the "cops in our
heads" while "the anarchic situation is *amoral*." "Morality is a form of
authority. . . . Morality and judgement go hand in hand. Criticism . . .
is essential to honing our rebellious analysis and practice, but judgement needs to be utterly eradicated. Judgement categorizes people as
guilty or not guilty—and guilt is one of the most powerful weapons of
repression" (73). Primitivist discourses attribute this repression directly
to the influence of civilization, and so their discourses about freedom,
naturalness, and so on, can be read as knowing or unknowing parables
of innocence (as Rousseau's *Discourse* also was).

Of course, while it is relatively easy to find examples of cultures
that lack the political trappings of modern civilization, such as a centralized state, and while as Arendt argues, politics—understood as a
social achievement, not an essential aspect of human interactions—is
often absent or at least ephemeral, there are no societies lacking some
form of morality. There are no innocent (amoral) peoples. Ironically,
those who are supposed to come closest to such a situation are usually societies that have been uprooted, shattered, and dislocated by
the effects of modernization (Turnbull 1972),[15] which only serves to
reveal the complex realities underlying primitivists' oversimplified
generalizations. And so, while the condemnation of oppressive moral
codes is clearly liberating, as is the rejection of political oppression,
the idea that oppression ceases in the absence of ethics and politics *as
such* seems entirely mistaken. To the extent that primitivists suggest
this, they might be thought of as surrendering the ethical and political possibilities for human life through their acceptance of what is
almost an inverted Hegelianism, a historicist view of civilization's
inevitable destructive end and of the restoration of nature's sovereign
authority.

Things, though, are not quite so simple because there may be other
potential practices that are neither political nor ethical but where
human freedom might still be envisaged as being exercised even in a
state of nature. This is why the notion of a future primitive condition
raises exactly the kinds of questions Bataille posed to Kojéve concerning Hegel's understanding of what happens to humanity at the end of
history, that is, when the dialectical movements of ethics, politics, and
philosophy have all finally been concluded (see chapter 2). As Agamben
notes, Kojéve suggests that after history's end,

man remains alive as animal in *harmony* with nature or given Being. What disappears is Man properly so called . . . the definitive annihilation of Man properly so called or the free and historical Individual—means quite simply the cessation of Action in the strong sense of the term. Practically, this means the disappearance of wars and bloody revolutions. And the disappearance of *Philosophy,* for since Man no longer changes himself essentially, there is no longer any reason to change the (true) principles which are at the basis of his knowledge of the World and of himself. But all the rest can be preserved indefinitely; art, love, play, etc., etc.; in short everything that makes Man *happy.* (Kojéve in Agamben 2004, 6)

This description, in many respects so close to those of primitivists, also illustrates what might be lost by surrendering (giving up on) politics, philosophy, and ethics as such. Because the cessation of "Action" (of politics in both Hegelian and Arendtian senses), of philosophy (for example, the "weakening" thinking that would continue to critique all principles claiming to encapsulate absolute, essential, or final truths), and the loss of individuals in their singularity (so vital for ethics) are associated by Hegel and Kojéve not just with the end of Man in terms of a metaphysical idea(l) but of individual freedom and the chance to initiate different possibilities.

As Agamben points out, the debate between Bataille and Kojéve concerns precisely this "rest," the residual practices supposedly still left at the end of history—"art, love, play, etc., etc."—because these too seem to require a certain initiating power, an ability to create change, to freely open new possibilities through what Bataille refers to as a "negativity with no use" (Bataille in Agamben 2004, 7). "Negativity" here has the sense of going against the established order of things, of contradicting the dictates of history or nature. "No use" refers to these contradictions escaping or exceeding the (totalizing) compulsion of a purportedly global Hegelian dialectic that turns everything (whether thesis or antithesis) into a means toward the end, the completion (final synthesis) of history. That is, these practices are of no use within and "reside" beyond (even continuing after the end of) the restricted economy of the Hegelian dialectic. But this "negativity with no use," this ability to freely initiate change, means for Bataille that humans (contra Kojéve) are not simply reduced to animality at the end of history; rather their condition is indeterminate, neither animal nor (distinctively) human. Their status mirrors, though Bataille does not make this connection, those regarded as the "incompletely human" precursors of

Lascaux's cave artists. Bataille imagined this future/primitive undecidable human/animal in terms of the amoral figure of the headless
man—Acéphale—"a being who makes me laugh because he is headless . . . [and] fills me with dread because he is made of *innocence and
crime*" (Bataille in Kendall 2007, 129, my emphasis)[16]—a figure that in
Bataille's own words is both mythic and monstrous.

Agamben agrees with Bataille's critique of Kojéve but wants to regard the "residents" of a posthistoric situation, whose human/animal
status is undecidable, in a rather different way, and this is precisely because he (like Arendt, and unlike Hegel and Bataille) regards politics,
and not (just) art, love, or play, as "pure means," that is, as a means
without (not teleologically justified by any predetermined) end. For
Agamben, politics provides a (indeed *the*) space for the exercise of individual freedom, since, following Arendt's argument, it is our acting into
the political sphere that both initiates change and reveals *who,* as an
individual, rather than *what,* we are. The idea of a world without the
possibilities of political action (one where politics has come to an end)
is not something to be celebrated or an idea(l) to direct one's attentions
toward. On the contrary, insofar as these possibilities are denied to
human individuals, they are reduced not to animality (as Kojéve suggests) nor to an acephalic monstrosity (Bataille) but to what Agamben
terms a state of "bare life," that is, of *human existence stripped of its
political possibilities and freedoms,* its abilities to initiate change in
concert, though not necessarily agreement, with others. This condition of bare life is also precisely that produced by the anthropological
machine when used to decide whether certain people(s) are, or are not,
properly or essentially human and thus whether they are forcibly exempted from any political community.

This apolitical condition is, Agamben argues, dreadful but, contra
Bataille, is hardly a laughing matter. Any persons reduced to bare life
can do nothing in concert with others to change this condition, for they
are, by definition, bereft of and exempted from political possibilities,
at the mercy of now incontestable (sovereign) powers. This remains
the case whether or not these powers are thought of as consequences of
historical or natural laws or, as Bataille (and many primitivists) might
prefer to think, of each individual giving in to his or her immediate
desires and needs—whatever the political and ethical consequences
might be. Exercising such a "freedom" is only "a sovereign and useless form of negativity" (Agamben 1998, 62) because it does nothing

to constitute a community with others or provide an understanding of oneself in relation to them. It might be more accurate to say that the only kind of community it could constitute is one like Acéphale itself, a "small group of forty year old initiates—who were not afraid to challenge the ridiculous by practicing 'joy in the face of death' in the woods on the outskirts of Paris, nor later, in full European crisis, to play at being 'sorcerers apprentices' preaching the people's return to the 'old house of myth'" (Agamben 2004, 6–7). To the extent that primitivism portrays the future primitive as a real option, it too is in danger of replaying ridiculous myths, acting out rituals in the woods while a full worldwide ecological crisis that requires political attention and action emerges.

To the extent that he emphasizes politics as *the* space of potential freedom, Agamben tends to overlook the possibilities of art, love, play, etc., just as Bataille, in his description of both prehistory and posthistory, emphasizes these very practices and their transgressive possibilities at the expense of politics as such.[17] This has important implications for distinguishing Bataille's and Agamben's very different understanding of the place and role of sovereignty. Put simply, Bataille seeks freedom through the figuration of an apolitical, amoral, acephalous sovereign being; Agamben, through a political (endless) critique of sovereignty. Bataille celebrates myth, Agamben seeks to dissipate it. Bataille is, as Agamben (1998, 112) remarks, "exemplary" in thinking about the nature of this being beyond politics, in showing how a human existence lacking political possibilities is still not reducible to animality. But the mistake he makes is to elevate the "radical experience" of bare life to a "sovereign figure" (112) inscribed in myth and in the interiority of its own transgressive experiences rather than seeing that this abject figure is actually the key to understanding the operation of both (political) sovereignty as such and the constitutive possibilities of political communities that might no longer be reliant on the inclusive/ exclusive definitions of the anthropological machine.

From the point of view of ethics, too, Bataille offers important insights, but again these are not necessarily those he intended. The reality, the "open wound," of his life in terms of his determination to transgress moral codes, to become as close to the figure of Acéphale as humanly possible, is no utopian vision of a return to primal innocence—as primitivism views posthistory. Such would only be possible if humanity was indeed, as Kojéve suggests, able to be reduced to animality (which itself presumes a questionable anthropological understanding of animals as essentially amoral beings). But Acéphale is, as Bataille's writings and practices show, an admixture of amorality and immorality, of "innocence and crime," of the "pure and filthy, repugnant and fascinating" (Agamben 1998, 112). Bataille is obsessed by transgression, and yet transgression only makes sense if there remains something, a moral code or, much more radically, an understanding of ethics as such, to transgress. This too is Kierkegaard's point. Transgression *cannot* be entirely innocent nor innocence (amorality) merely a matter of transgression. Anyone with an inkling of the Good, even those like Bataille who attempt to ex(or)cise it, will therefore inevitably recognize a taint of immorality in their desire to transgress that sense of responsibility for Others, which is, in Levinas's sense, already beyond Being.

The return to a pure amorality (a state of innocence) would not then be just a matter of overcoming the "cops in our head," of ceasing to mindlessly follow moral norms, but of attempting to expunge ethics as such, that is, to eradicate all understanding of and feelings concerning that matrix of responsibility in which, according to Levinas, we initially find ourselves in and through loving relations to others. Stripping away layers of social mores and expelling all selfless concerns for others would neither reveal an original (self-enclosed, sovereign, autonomous, free) individual nor release some culturally repressed primal animal nature. Its result would be the ethical equivalent of a bare life, a life divorced from all concerned involvement with others as Others—not so much headless (acephalic) as heartless. And this is not just politically and ethically destructive of the possibilities of living together in communities with other beings (human and more-than-human) but, quite literally, self-destructive. Indeed, the desire for absolute transgression (to prove one's absolute freedom from ethics) can only reach its ultimate limit, be resolved, in death and not in life with others (ethics and politics): only "through the instantaneous transgression of [the ultimate ethical/moral] prohibitions on killing" (Agamben 1998, 113) or by a permanent resolution in the obsessed subject's own death.[18]

Innocence is not something that can be recuperated: once recognized, it is already lost; once lost, it is gone forever. Anyone capable of exercising a sense of responsibility for others as Others cannot claim to inhabit a state of innocence, to be entirely amoral. To be innocent (amoral) is a matter of the absolute absence of ethical responsibility and of any experience of ethics. One can be deemed more or less innocent

concerning certain ethically questionable actions (and ethics as such is
this putting actions into question), but this is something quite different,
since this kind of innocence depends entirely on possessing an ability
to exercise ethical judgment, not on the absence of judgment. Contra
Feral Faun, being able and willing to make such judgments for oneself
is the sign of one's taking ethical responsibility, a sign of one's individuality and of a desire not to simply accept the present order of things as
if it were incontrovertibly "natural." Arendt makes just this point in
relation to those who resisted the sudden restructuring of moral norms
in Nazi Germany because they considered them immoral. Exercising
their own ethical judgment (and their sense of responsibility), these
individuals opposed what now passed for normality even though doing
so placed them in life-threatening conflict with the claims of the new
"moral" authority.

To take the idea(l) of innocence associated with the state of nature
literally and use it to deny all ethical responsibilities is to espouse a
form of amoral absolutism. That is, it advocates a political principle
of theological antiauthoritarianism that awards sovereign individuals
the authority to grant themselves absolution from their damaging effects on others, to wipe clean the slate of their souls: it celebrates an
ethically bare life. In this too it simply reverses the usual political order,
which is based in institutions' self-awarded authority to absolve others'
"sins." After all, the idea of being able to grant a return to a state of
innocence, to absolve guilt, is actually the most powerful, most fundamental, and the original (theological) basis of repressive political and
moral institutions, of church and state!

What sense can there be, then, in primitivism's appeal to innocence?
It is certainly not an end (a reality) that can be reached, not even one to
be desired in any utopian fashion. But, as already intimated, the role
of innocence in a future primitive state might instead be regarded as a
philosophical myth deployed with ethical (and political) intent (just as
Plato's myth of the cave can be read as deploying an ideal of the Good
with ethical intent, as a metaphysical guide to ethics in Murdoch's
weak sense rather than as instituting a strongly metaphysical, authoritative claim). In this weak(ening) sense, the ideal of innocence has a
certain negative capability as a means to encourage the exercise of
self-critical judgments about the current condition of the world. When
coupled with other traits associated with the state of nature, especially
its presumed "wildness," it also has a life-affirming intent insofar as it

suggests a further source of initiating freedom, that, unlike notions of "art, love, play, etc., etc.," goes far beyond the categorizing claims of the anthropological machine. That is, wildness offers the possibility of a "negativity with no use" that is by no means confined to a specifically human existence, one that escapes and challenges all forms of restrictive economy. In this sense, when understood as a myth with intent rather than a real possibility, primitivists' evocation of the state of nature and the future primitive contributes to the continuance of (ecological) politics as pure means and of (ecological) ethics in terms of its concerns with others as impure ends, as mortal beings (see chapter 2). It does this by beginning to reevaluate and criticize the fundamental, orthodox myth of the origins of modern political and ethical authority.

But how does an idea of innocence operate in this way? It is made to appear in mythic scenarios that illustrate how the moral absolutism underlying contemporary claims to political authority might be weakened.[19] Specifically, this mythic dimension is counterposed to a culture of contamination and guilt insofar as these foundational myths too recognize that innocence is far from being a civilized virtue. As an original condition—a myth and idea(l)—innocence is opposed to progress and civilization in almost every way. The innocent wanderers in the state of nature had no call to consider the propriety of their picking fruit, nor were they yet corrupted by avarice or made subject to the authority of any moral order, only the "laws" of an amoral nature. As innocents, they were also unable (or unwilling) to consent to contractual obligations with others. Innocence is then an ideal that, however envisaged, remains disassociated from the world of use, exchange, and even moral values, something that lies outside but is (as Rousseau too argued) inevitably consumed by contact with civilization.

Modern Western civilization accepts this loss of innocence but in a particular way: by declaring that because of a kind of species-wide (anthropologically determined) hereditary sin, all humans are equally guilty of accepting this loss of innocence. We are all, as the theological model purports, descended from Adam and Eve. We are all, as its incompletely secularized Enlightenment progeny supposes, subject to our human nature. So the story goes—we are then *all* guilty of plotting to leave the state of nature, of desiring and competing for powers over others and for private property, of avarice concerning the accumulation of the only incorruptible material in the world (money), of needing and wanting to be subject to an overarching moral order, and of sign-

ing a political contract with authoritative institutions in the name of
personal (and financial) security. The subjection of the world and all its
inhabitants to the authority of markets and states is portrayed as the
inevitable result of everyone's self-serving desires, as everyone's fault,
but also of our accepting the moral and political authority of the state
as the very condition of the possibility of civilization.

But, when counterposed to a heterodox reappropriation of prehistory, the totalizing claims of this ethicopolitical surety begin to
break open. It is revealed, as if for the first time, as only a myth, with
no original (prehistorical) basis in reality, a myth that can be subjected
to ethical and political critique in its entirety. And a vital part of this
critique is to challenge this notion of civilization's original sin. This is
the ethical and political intent of innocence, not to offer an actual way
of assuaging everyone's guilt but to reveal the importance of resisting
the claim that everyone is *always* and *equally* guilty—guilty *in a way
that divorces the idea of ethical responsibility from any particular
acts, that systematizes and disperses guilt as a principle of ethicopolitical authority.* And while the amoral absolutism (there is no guilt)
of primitivism suggests an impossible alternative to a culture of contamination where all are systematically made guilty, when not taken
literally, its intent is to get others to think about how the structures
of authority that institutionalize such destruction operate and about
where authority's claims about responsibility lie (falsify things such as
they are) concerning the destructive inevitability of our current situation and where real responsibilities lie (might be found).

To follow this critique further, to give it ethical and political intent, requires exercising (not exorcising) the capacity to make ethical
judgments for oneself. Otherwise, primitivism simply replicates the
strongly metaphysical aspects of modernism's mythology, though in
an antiauthoritarian and noncelebratory way, by holding everyone
and every aspect of civilization equally guilty. But this is ethically and
politically useless, since, as Arendt argues, where everyone stands accused en masse, or where the whole process of history is to blame, as
for example, all Germans and/or German history were often made responsible for the emergence of Nazi Germany, this "in practice turned
into a highly effective whitewash of all those who had actually done
something, for where all are guilty no one is" (2003, 21; Agamben 1999).
Insofar as the modernist myth of the state of nature disperses guilt systematically, this only serves to deflect guilt from those most responsible

for bringing about and profiting from this condition. Primitivism, taken literally, does exactly the same. But when this myth is understood illustratively and critically in terms of its ethicopolitical intent, it reveals the complicity between those in authority, who offer, and those subject to authority, who seek, absolution and the ways in which this serves the purposes of the current ecologically and socially destructive system.

This talk of the complicity between absolute innocence and systematic guilt may seem a little abstract. But think, for example, of the ways in which we are frequently told by authority that the ecological crisis is everyone's and/or no one's fault and that it is in all of our own (selfish) interests to do our bit to ensure the world's future. Think too of the tokenistic apolitical "solutions" this perspective engenders—let's all drive a few miles less, all use energy-efficient light-bulbs, and so on. These actions may offer ways of further absolving an already systematically dispersed guilt, but they hardly touch the systemic nature of the problem, and they certainly do not identify those who profit most from this situation. This pattern is repeated in almost every aspect of modern existence. Think of modern cityscapes or of the shopping mall as expressions of modern civilization's social, economic, and (im)moral orders. These are far from being places of free association. They are constantly and continuously monitored by technology's eyes in the service of states and corporations (Lyons 2001). Of course, those who police the populace argue that the innocent have nothing to fear from even the most intrusive forms of public surveillance, that it is only the guilty who should be concerned. But this is simply not true, nor, as Foucault (1991) argued in a different context, is this the rationale behind such panopticism. Everyone is *captured* on closed-circuit television, and it is precisely any semblance of innocence that is lost through this incessant observation of the quotidian. All are deemed (potentially) guilty and expected to internalize the moral norms such surveillance imposes, to police themselves to ensure security for private property, the circulation of capital, and fictitious (anti)social contracts—"No Loitering Allowed," "Free Parking for Customers Only." This egalitarian dispersal of guilt contaminates everyone, placing them on interminable trial, since no one can ever be proven innocent by observations that will proceed into an indefinite future.[20] Thus, as Camus (1984, 12) remarked, it is indeed innocence that is called upon to justify itself, to justify why it should (but will not) be allowed to survive in any aspect of everyday lives.

This surveillance is also (and by no means accidentally), as Agamben argues, a key aspect of the biopolitical reduction of politics to the policing of disciplined subjects, redefined not as individuals but as "bare life." Again, this refers to people stripped of their political and ethical possibilities and now primarily identified in terms of their bodily inscription of transferable information. People are not quite reduced to animality, to *just* their biology, but their biological being is made increasingly subject to observation, management, and control as the key mode of operation of contemporary authority. Video cameras; facial, gait, and voice pattern recognition technologies; fingerprints; retinal scans; DNA analysis; electronic tagging; the collection of consumer information; data mining and tracking; global positioning systems; communications intelligence; and so on, concern themselves with every aspect of people's lives but are in no sense concerned for the individuals (in their singularity). Rather, they measure and evaluate that person's every move as a potential risk to the security of property, the security of capital(ism), the security of the moral order, and the security of the state. The entire populace comes to occupy an increasingly pervasive state of exception, a "zone of anomie" (Agamben 2005, 50) that is certainly not a state of nature but a technologically mediated political (and ethical) void. And again, if this authoritarian monitoring and control is questioned, the answer is that it is everyone's fault and nobody's, that it is actually our desires and our ultimate personal security (the security of people and state) that determined that the relevant authorities had no choice but to take this path, that it is a small cost to pay for the protection of civilization, that, in effect, we are all guilty of our own impending technologically mediated reduction to bare life.

This kind of biopolitical reduction of society to managerial surveillance and control is the very opposite of anarcho-primitivism's understanding of a future primitive world without authority where humanity is reembedded in nature. And yet, these two extreme prospects are both, in almost every respect, fundamentally premised on the utopian/dystopian reductions of ethical and political possibilities to forms of bare life. One is based on personal risk rather than collective security, on political anarchy rather than political authority, on falling back into nature's cyclical patterns rather than civilization's progressive (linear) transcendence of nature. One is a political impossibility, the other an increasingly apolitical reality. If completed, both would, albeit in very different ways, put a stop to the anthropological machine (but only at the cost of putting an end to ethics and politics). Primitivism does this by supposedly eradicating all cultural alienation from nature, expunging all those cultural features that set humans apart from nature (even, in some cases, including the use of language, which Zerzan [1999, 31] has described as "the fundamental ideology").[21] Progress, on the other hand, threatens to achieve this through the reduction of real people to technologically mediated and engineered, biologically encoded, and bureaucratically manipulated information—to cybernetic organisms, systems within systems. But one does not have to be a humanist (a proponent of the anthropological machine) to find both of these options ethically and politically offensive. And so, what is required are ways of stopping the anthropological machine without losing the vitally important and varied forms of negativity with no use that keep open the ethical and political texture of life (see chapter 4).

Wildness

Focusing on the ethical and political intent of primitivism helps elucidate another, quite different form of negativity with no use, the liberating and life-affirming potential that is often rejected by humanists precisely because of its anticivilizational connotations, namely, that of wildness/wilderness. Indeed, primitivism, deep ecology, and radical ecology all to some degree exemplify an ethos of wild(er)ness. To speak of an ethos in such circumstances, especially given amoral (literal) readings of the state of nature, might seem strange, yet this ethical sensibility is readily apparent. Unsurprisingly, it goes against the grain of both the specific values that have dominated modernity and the dominant ways of formulating ethical concerns, that is, as abstract and supposedly universally applicable values (rights theories), methods (Kant's categorical imperative, Habermas's ideal speech situation), and/or calculative systems (utilitarianism)—see Smith (2001a, 2007b). It is ethically anarchic. The values espoused and the forms taken by such ethical values recognize experiences of wildness as an inspirational (animating) source of individual freedom, a wildness that rejects all attempts to impose a civilizing moral order.

In common with many elements of deep and radical ecology, primitivists celebrate what they regard as the wildness of an unconstrained and untrammeled nature, of an unexploited world not yet reduced to use and exchange values, that is neither fully commodified nor domesticated. For example, they support the activities of those associated with Earth First! who are concerned to "fight in defense of wild living beings and places that haven't yet been destroyed" (Black and Green Network n.d., 1). In many cases, they also link together ecocentric critiques of instrumental understandings of nature with ecofeminist critiques and are critical of the ways in which the anarchist tradition itself has "been silent, in many ways about the domination of animals and nature, and the connections between them and the suppression of the female or feminine principle, by patriarchy" (1)—anarchy has not, after all, contra Morris, always implied an ecological attitude.

Of course, it might be objected that the very idea of wilderness as civilization's Other, as something to be saved from human encroachment, is a historically particular product of modernity. An extensive literature tracks changing understandings of nature (for example, Castree and Braun 2001; Cronon 1995), including specific historical work on changing ideas and evaluations of wilderness and on primitivism itself (McGregor 1988; see also Nash 1982, 47–48). Without entering discussions about the social construction of nature here (but see Smith 1999a), such literature would clearly suggest that primitivism too has a history, that its interpretations of nature and notions of self-liberation do not just arise instinctively or emerge from unmediated exposure to wilderness—however defined. Rather, they call on cultural precedents and rely on a "metaphysical vocabulary" to use Soper's (1995, 61) phrase, "that has developed in tandem with the development of Western culture or 'civilization' itself." However, while primitivists (and deep ecologists) often use the term *wilderness* rather uncritically, and while little if any of nature remains in a pure, uncontaminated state, there are clearly places that are relatively wild in the sense of not being under constant human surveillance, regulation, and control, where nonhuman life continues relatively unhindered. Such places can and do have profound effects on individuals' self-understandings and values.

It also needs to be stressed that unlike deep and radical ecologists, primitivists tend to place emphasis on wildness rather than wilderness per se. As Zerzan (1994, 146) puts it, "Radical environmentalists appreciate that the turning of national forests into tree farms is merely part of the overall project that also seeks their own suppression. But they will have to seek the wild everywhere rather than merely in wilderness as a separate preserve." Wilderness is certainly regarded as a source of wildness, a setting within which an upwelling of initiating events that resist being completely tamed and controlled are expressed, but it is not the only setting in which this can happen. In short, wildness is regarded as synonymous with creative freedom from social constraint, as that excess that escapes the controlling dialectics of civilization. In this sense, deep ecological understandings of wilderness come close to but do not provide the complete story. From the perspective of primitivism, "wilderness experiences" allow a person to recognize and express something of the wildness that, insofar as they remain natural, living, beings, lies within all people. It both illuminates humanity's affinity with wild nature and, in freeing individuals from the social obligations of everyday life, reveals the extent to which their lives have become dominated by the social necessities of progress. The domestication, control, and suppression of wild (free) nature and individual human freedoms are thereby related directly to the same civilizing process—both are made subject to increasingly pervasive social techniques and forces.

For primitivism then, "anarchism implies and incorporates an ecological attitude towards nature" (Morris 1996, 58) in a fairly direct way. Anarchism, understood as freedom from constraint, *is* wildness, and that wildness is the creative, living aspect of nature, both wild nature and that original human nature now dominated and repressed by the civilizing process. And if primitivists' espousal of such values have "irrational," "mystical," and mythic overtones, then these are certainly not those associated with organized religion, hierarchic structures, or the imposition of a fixed moral order. Indeed, there is more than an echo here of what Vaneigem (1994) refers to as "the movement of the free-spirit," the recurring exuberant tendencies, like the Beghards and the Beguines found throughout the European middle ages, which rejected all religious institutional authority and all attempts to impose a moral order on individuals by church or state. These tendencies too often involved Edenic narratives of lost innocence. Vaneigem's (1994, 132) description of Margaret Porete, burned for heresy in Paris in 1310 with her book *The Mirror for Simple Souls,* might almost describe contemporary primitivism. "Whereas the Church was hostile to nature, Margaret proposed that it might be rehabilitated to its state before the Fall, before the appearance of sin and the invention of exchange.

Access to it lies in the refinement of love, in an identification with the power it generates in everyone ... desires are awakened and pursued freely, in total innocence, and with no sense of guilt." The concern here is one of freely given love rather than wildness, but the message is remarkably similar, of reconciliation with a natural state of innocence and the attainment of liberty through recognizing and releasing the anarchic free spirit embodied in everyone.

Another aspect of value to be drawn from primitivism, then, might be that wildness too is a creative vitality initiating otherwise un-predictable actions and unexpected turns of events; it is a life-affirming negativity with no use that resists totalizing attempts to impose au-thority and order on life itself. And this vitality cannot be reduced by the wiles of the anthropological machine to a sign of human ex-emption, as ethics, politics, art, love, and even play have sometimes been, precisely because it epitomizes the more-than-human world that civilization seeks to exclude, destroy, and tame. Wildness is how we understand that nature too resists being totally encompassed in and domesticated by modernity's controlling schemes. And in this sense, since wildness is something also recognized and valued in some aspects of individual human lives (for example, it has an affinity with Arendt's [1958, 178][22] description of the natality of political action [Schell 2002] and its unexpected consequences—see earlier discussion), it also offers a possibility of breaking the hold of the anthropological machine.

This, of course, raises another question: If the various forms of nega-tivity with no use noted by Bataille (art, love, play, etc.) and Agamben (politics) signify how human beings can express creative possibilities (possibilities that can also be stripped from or denied them), what does the wildness of the more-than-human signify? Perhaps it can signify the initiating, creative, natality of nature, its unpredictable adaptabil-ity, its giving rise to worldly singularities and ecological diversity, and its resistance to being reduced to those totalizing systems, concepts, laws, processes, and especially to anthropocentric use and exchange values, which assume an authority to organize, manage, and control nature—it signifies how life exceeds the boundaries and categories of such systematic impositions, how nature is never just a resource. Wildness, in this sense, is not something that can be captured by sci-entific description, though it might sometimes be glimpsed through the lens of scientific insights. But more than this, wildness signifies that life on Earth is not just a matter of survival, that it is not just the ongoing repetition of the Same, but also a creative and potentially infinite movement of irreducible differences.[23] Wildness is not something isolable or locatable—a point of origin that can be defined and controlled (although it is always instantiated in particular events and places)—but an opening that escapes the order of things authority always seeks to impose. Wildness is an unspecifiable expression of that worldly singularity of beings that an ecological ethics concerns itself with, experienced and gathered together with necessarily incomplete understandings of our being-with-others. It is what shatters the predictable patterns of everyday routines in experiencing the beauty and fragility of mortal existence and sensing what it feels like to be alive at those moments when things seem most intense, when we weep or laugh, touch ice-cold water or the hand of a loved one, at those moments when we are momentarily diverted from our self-interested obsessions by the sight of the kestrel's hovering flight. *Wildness is the initiating and differentiating creativity, the life-affirming, animating negativity with no use, the natality of beings, by which the more-than-human world resists dominion and its reduction to a resource.*

This contrast between life and survival is a key aspect of the ethos of primitivism and one that it shares with many deep ecologists and radical environmentalists. Primitivism envisages civilization as nothing more than a survival machine with absolutely no benefits. From this perspective, even the great success stories of modernity, like modern medicine, increasing longevity, and so on, all come with a price attached—our increasing dependence on the machine—and all are supposedly in the process of unraveling as, for example, drug resistance increases and ecological disasters loom. The machine offers an illusion of control over the world, but from primitivism's perspective, each turn of the wheel actually decreases the individual's potential to experience life in its fullness as we become ever more dependent, domesticated, and alienated from our own vital potential and the living world. Willingly or unwillingly, the civilizing process ensures that we come to trade life and liberty for mere survival-regulated continuance over time. Agamben's concerns about modernity's technological and bureaucratic reduction of human beings to bare life unknowingly echo this very sentiment.

While progress advertises itself as an adventurous and advantageous journey into an unspecified future—and thereby seems to hold open some space for technologically mediated creativity—it does so primarily in ways motivated by profit and the insatiable desires and dictates of a modern market system entirely dependent on the reduction of the world to use and exchange values, to a resource. The binding contracts it presumes already signed trade the real possibilities of freedom that the natural world offers in the here and now for the imaginary security guaranteed by a new Leviathan, an authority based in an otherworldly imaginary that increasingly reduces real people to bare life and nature to an expendable matter of calculated risks and profits—or at best to that managed environment necessary for ensuring the survival of this same economic system and of bare life (assuming, of course, that modern civilization's inventions and interventions in the form of nuclear weapons or global ecological disasters do not eradicate even this hopeless possibility).[24]

The myths that primitivism invokes are not just fables for the foolish but express a critical ethical and political intent—although this appears only when such myths are not taken literally as calling for a reversion to some precivilizational ideal. It would be far from liberating to reject the potential freedoms of politics and ethics as such in order to save something called wildness. It would also make little ecological sense, since the particular wildernesses that are the foci of much environmental campaigning are in far more immediate danger than wildness per se. Yet the promise and the problems of any relation between ecology and anarchic ethics and politics coalesce around this issue of wildness. The promise lies in realizing ways to connect concerns about human well-being and liberty to the more-than-human ecology of the world, ways that would go beyond the boundaries imposed by the anthropocentric humanism that characterizes shallow ecological approaches. Such an approach might make real contributions to environmentalism in general and radical and deep ecology in particular, opening debates that exceed their current confines. After all, many, perhaps most, environmentalists find wilderness special because, and to the extent that, it is wild. They deem it valuable because it is not entirely dominated, monitored, transformed, and constrained or made to conform to the dictates of its efficient utilization by humans. To be sure, most recognize that wildness can be found in many places, from the pavement-cracking weeds on city streets to those moments of liberating rebelliousness within us all. But wildness should also exist as something big enough to lose and find oneself in, something that draws us out of ourselves and the narcissistic culture that Bookchin so wrongly regards ecological anarchy as party to. In this sense, we are not talking about life-style anarchism but a living (ecological) anarchy.[25]

However, the problems of any anarchic ecology also concern wildness: first, in terms of whether this idea can be conjoined with a coherent account of modern civilization's ills that avoids the all-or-nothing fundamentalism of literal interpretations; second, in recognizing that an idea of wildness, while necessary, will not by any means be sufficient in and of itself for understanding the free existence of people in relation to each other and to nature. Any real understanding of people, freedom, or more-than-human beings is diminished, if not impossible, without ethics, politics, and all the other myriad dimensions of art, love, play, etc., etc. Third, and closely related to this second point, purist versions of primitivism fail to recognize that the selves they want to liberate, no less than the nature in which this liberation is to occur, are themselves socially and historically entangled and composed. One cannot reject society and history en masse and still retain those self-understandings that have their sources within that same "civilizing process" (Taylor 1996).

This is a point that Bookchin (1995a, 14–15) also made, and such a rejection is indicative of a lack of self-understanding, that is, a lack of understanding what it might mean to be a human self, which always, in all times and places, involves being part of a culture. Not one of the peoples that are called upon as exemplars of primitivism's preciviliza-tional ideal exists in a state of nature. Not a single person in all these peoples came to their own very different understandings of what being who they are involves through simple exposure to wild(er)ness. Rather, they all found themselves (were immersed and came to self-awareness) through their extraordinarily variable experiences of social as well as natural situations. Bearing these complex historical interrelations among selves, societies, and natures in mind, and explicitly thinking through them (in however many possible ways), will be vitally important in creating a genuinely different understanding of wild(er)ness and its political import.

In this sense too, wild(er)ness might become important for other understandings of politics and ethics that, while not explicitly anarchist, still regard these in terms of a negativity with no use and as open textures (see chapter 5). The challenge that primitivism poses is one of diversifying and deepening the myriad possible connections among ecology, politics, and ethics in terms of an anarchic understanding of wild(er)ness. For, as Thoreau, that most influential of environmental philosophers, famously declared, "In Wildness is the preservation of the world" (1946, 672). And Thoreau's point, too, as he explicitly said, was to make an emphatic and "extreme statement," for he believed there were already "enough champions of civilization" (660). Instead, he chose to "speak a word for Nature, for absolute freedom and wildness, as contrasted with freedom and culture merely civil" (659). This is an aim that many might share.

4 SUSPENDED ANIMATION

Radical Ecology, Sovereign Powers, and
Saving the (Natural) World

How, then, to "speak a word for nature," as radical
ecology tries to do, if this speaking is now reimagined as a critique of
both the principle of sovereignty and the divisive operations of the anthropological machine? How might the ethical and political concerns
of radical ecologists to "save the world" actually be expressed in terms
that voice the anarchic aspects of ethics and politics, and what kinds
of theoretical constellations might help guide such an endeavor? What,
indeed, might saving the world mean?

instrumently—for us.

ethicly—for itself

From Standing Reserve to Saving the World

The idea of saving the (natural) world has about it an air of ridiculous naivety. First, it seems unrealistically grandiose in the scope of its
ambition. How could one hope to save a whole world or to keep all of
nature safe? Second, it appears too close to the patronizing and dangerous religiosity of those who want to save America or our souls for
Jesus and free enterprise (a somewhat strange combination), whether
or not we want to be so saved. Does the natural world, which is, as
Murdoch remarks (see chapter 2), so "alien," ultimately "pointless," and
"independent," really want or need saving, and for whom? Third, it is
all too readily compared, and all too rarely contrasted, with the kind of
mindless fundamentalisms that, with proselytizing fervor, posit single,
simple, but mutually contradictory ends for humankind. After all, are
there not many worldviews and correspondingly many understandings
of what saving the natural world might entail? Of course there are.
And yet it might still be suggested that, deep down, radical ecologists

strive to save what they can of the natural world, that this is their fundamental ethical and political concern.

What is more, this ethical and political concern separates radical ecologists, those who would go to the root of that which threatens the world, from the purveyors of environmental expediency, from the "shallow" (to use Arne Naess's [1972] term) environmentalists who formulate all concerns for the natural world within the globally dominant language of resource economics and management. It expresses the difference between those who regard the natural world as a realm of impure ends, as beings of indefinable (infinite) value but finite worldly existence (see chapter 2) and those who consider it merely a "storehouse of means," of value only because of its potential usefulness toward humanly determined needs and desires. On this latter view, the world is worth saving only in the sense that one might prudently save money for a rainy day, only as "natural capital" that earns us interest, rather than as that which is deserving *of* our interest, our concerns.

Radical ecologists, then, argue that a distinction between ethics and instrumentality is no less important with regard to the natural than to the social world. Saving the natural world is an ethical end in itself. But, what kind of an end can it be? In what sense can we speak of a natural world of ends, and how might this be related to concerns about an ecological crisis, that is, the potential ending of the (natural) world? Does the rejection of resourcism not confirm that radical ecology is, as skeptics suppose, just another ridiculous form of fundamentalism (like primitivism) naively refusing to engage in realpolitik? There are no simple answers to such questions, although just admitting this already begins to distinguish radical ecology from any single-minded fundamentalism. An initial step might be to distinguish a realpolitik that provides a systematically applied excuse to compromise one's ethics from a "politics for the real (natural) world," understood as an applied art of seeking, wherever possible, ethical "compromises," that is, a worldly *phronesis,* an ethically inspired political wisdom.

Of course, speaking the language of resource economics may, on occasion, persuade sovereign powers to grant this or that aspect of the natural world a temporary stay of execution. But, as Neil Evernden (1999) argues, it also, wittingly or unwittingly, accepts the original terms on which nature's death warrant has already been signed. It concedes everything to an understanding of the world as no more than what Heidegger (1993a) calls a "standing reserve" of lifeless, that is, deanimated and nonautonomous "matter," systematically ordered according to a technological enframing *(Gestell)*.[1] The forests and their myriad inhabitants are thus conceptually reduced to so many board feet of timber, the once roaring rivers to so many kilowatt hours of hydroelectricity. From more radical perspectives, and at the risk of seeming ungrateful for small mercies, we might regard even those patches of the world momentarily set aside from more corrosive forms of technologically mediated commodification as beings left in a state of suspended animation, as hanging dearly onto bare life above the gallows-drop of global capitalism. This is the condition we have already noted in terms of the nature reserve (see Introduction), and the fate of the world's whales might offer another case in point—though *fate* is not the right word here, since their salvation or extinction is, for the moment at least, in human hands and not an issue predetermined by irresistible (super)natural forces.

We should bear in mind, then, that, like realpolitik, fate too provides a (historicist) rubric that falsely naturalizes worldly apathy. Both terms imply that ethicopolitical action is irretrievably subservient to sovereign powers (whether such powers are envisaged as progress or the invisible hand of the market), which we must simply accept because they cannot be resisted. But neither term has a place in a politics for the real (natural) world precisely because, at least from the perspective of radical ecology, this naturalization is false. Nature is not the source of the short-term, calculating, self-interested individualism that constitutes the (a)social world envisaged by contemporary advocates of realpolitik, nor should it be made subject to it. Nor is nature a synonym for, or ruled by, fate's decree; it is not governed by powers that impose a predetermined order on the world's unfolding. The radical ecologist does not want to save whales from realpolitik only to make them subject to some other predetermined fate (as those who reject all interference in natural processes might do), nor do they want to preserve them in timeless aspic in a museum or a dolphinarium. To save the whales is to *free* them from *all claims of human sovereignty,* to release them into their singularity, their being such as it is—whatever it is—*quodlibet ens,* and into flows of evolutionary time, of natural history, just as they release themselves into the flows of the world's oceans. This "saving" is an ethicopolitical action.

Of course, there is much more to say about this saving, but a politics for the real (natural) world must then recognize that the technological

enframing of the world, its ordering as standing reserve, its being conceived as merely an instrumental means to human ends, is not fated
either. Heidegger (1993a, 330) warns against "the talk we hear more frequently, to the effect that technology is the fate of our age, where 'fate'
means the inevitableness of an unalterable course." Nevertheless, while
not irresistible, this enframing is, in fact, the "supreme danger" from
which a politics of natural reality must strive to save us all: whales,
humans, indeed the whole world. For, at this very moment, when humanity "postures as lord of the earth" (332), it too risks being reduced
to standing reserve, to a material resource open to manipulation and
transformation.

How could this be? The systematic ordering of the world according to this technological enframing is usually taken as a sign both of
humanity's successful dominion over the external world and of the insuperable difference between human subjectivity and an objectively
understood natural world. But Evernden (1999), following Heidegger,
argues this view entirely misconstrues the nature of the world and of
human existence. Such objectification entails the rejection of our actual
phenomenal experiences of concerned involvement in the world, such
as the feelings of elation or freedom on a windswept mountaintop, or
despair and anger at the destruction of a well-loved place. We have to
be trained to regard nature objectively and dispassionately: seeing a
tree as protopulp for paper manufacturing is an "accomplishment," one
that requires us to overcome the childish notion that the natural world
is "alive to us." In other words, an understanding of the world as standing reserve has to overcome, to conquer, our phenomenological naivety.

This feeling oneself part of (which is not the same as feeling at one
with) a living (wild) world is not just the ground of radical ecology but
is expressed and made manifest in the phenomenal ground and flow
of every human existence. It is certainly a sign of the successful dominance of the technological *Gestell* that many "sophisticated" adults
claim to no longer feel this (or that they have managed to repress such
feelings) and that an entire polity is ordered on the basis that such feelings are unimportant. (The fact that this enframing has resulted in the
successful eradication of nonhuman natural beings from an increasing
proportion of so many human lives doubtless fosters this.) But this
Gestell should be seen for what it is: a bizarre historical aberration,
and one that, radical ecologists would argue, is closely connected to
our current ecological problems.

When the world is challenged to appear in this technological *Gestell,* when nature is set upon and set in order as a resource, then, for Heidegger, humanity too "stands within the essential realm of en-framing" (1993a, 329). And again, this is not at all to say, as we might be tempted to do, that the fate of humanity and the world are in-extricably entwined, because as already argued, this is not a matter of fate. Heidegger's ontology is, in any case, much more intimate and this-worldly, and much less determinate, than this. The enframing of the world encompasses human being *(Dasein)* because our existence is always already that of a being-in-the-world. The world only appears as it does through our being-*there,* our emplacement within it. Its ap-pearance as standing reserve is the expression of a particularly limited kind of human involvement within the world. To say that nature is a resource is to express something of that limited and limiting mode of existence. What it expresses is that we have forgotten the "nature" of our being and, we might add, our being in nature, forgotten that we can inhabit a living world of ends. What it now threatens is the end of the world as anything other than an ethical- and political-free trade zone, a profit-driven system of circulating resources.

If we regard the natural world as nothing but a resource, then hu-manity is left, at best, with nothing to become other than the orderer of that resource. At worst, human lives come to be entirely dictated by this projection, by our being caught up in endless cycles of resource mobi-lization. This is close to the reality of much of contemporary existence, where, to use Heidegger's example, the forester "is made subordinate to the orderability of cellulose" (1993a, 323). To view the world as stand-ing reserve, as a resource, then, is a dangerous self-fulfilling prophecy that is ultimately self-negating: it denies the natality and ethicopolitical autonomy of human being (of the self's existence). Paradoxically, the presentation of the world as just a means to suit human ends risks eroding the freedom to determine one's own destiny, to have one's own life unfold as an ethicopolitical end in itself in the company of others. Perhaps, ironically, it is only the fact that humans have the possibility of being-alive-to-the-world that offers any possibility of salvation here from the spiraling self-referentiality of economically driven "realities."

Humanity's posturing as sovereign lord of the Earth fosters an illu-sion that everything we "encounter exists only insofar as it is [humani-ty's] construct" (1993a, 332) and an accompanying delusion that we "al-ways and everywhere encounter" only ourselves (an illusion/delusion sometimes all too present in those writing about the social construction/ production of nature). The world-creating activities of nature are covered over, hidden from us as we come to consider everything of worldly significance a product of our own doing. We come to regard humanity as a world apart, somehow existing outside of the natural world, returning to it only to satisfy our socially determined needs. But we are not a world apart. Human existence is not, in any sense, ultimately separable from its existence in the world. We are beings that can only exist insofar as we stand out (ek-sist) into this world so that, as Heidegger emphasizes (332), we *"can never"* encounter only ourselves.

In a world where a technological enframing predominates, the place for ethics and politics (and also, art, love, play, etc. etc.) is correspondingly diminished, for they are ways of envisioning and creating a good life with those others we come to regard as being (to adopt the Kantian idiom) ends in themselves, those singular, indefinable, beings such as they are. Radical ecology, then, is contrary to its misanthropic portrayal by its many detractors: not just interested in saving the natural world, it is also a movement that strives to save a place for politics and ethics. For, one might say, *it is the reduction of the world to a standing reserve that threatens to reduce humans to the status of "bare life."* This threat is not just a dystopic possibility but, according to Agamben (2000, 36), already constitutes the "hidden matrix" of contemporary (bio)politics (see chapter 1). If Agamben emphasizes the human impact of the loss of political possibilities (of a political "negativity with no use"), a reading of Heidegger might suggest something more than this: that such reductive forms of biopolitical "inhumanity" become more likely when the world itself comes to be enframed as nothing more than a standing reserve. Unless we can think the roots of this technological *Gestell* (and then employ this thinking as a basis for political and ethical action), such biopolitical dangers will remain with us. (Though given Heidegger's association with, and failure to publicly repudiate, National Socialism [Farias 1989; Ott 1993], one might justly consider whether he reflected on his own role in this antipolitical and unethical enframing and its appalling consequences, a fact not lost on his own students, like Levinas, Arendt, Marcuse, and Löwith.)[2]

From a radical ecological perspective, we cannot save politics and ethics from this technological enframing without saving (wild) nature too. A weakness in Agamben's approach seems to be its failure to be concerned about the designation of the nonhuman world as being placed in a permanent state of exception (see later in this chapter).[3] As already argued (see chapter 3), we might say that the natural world is precisely where the state of exception originally takes the form of the rule, at least where dominant modern Western philosophical and political traditions are concerned. Nature enters politics and ethics primarily as that which ruling powers define their present political state *over* and *against,* as that apolitical realm that they first and foremost claim to exercise sovereign power over (as exemplified in Locke). The natural world is thereby reduced to property, resource, and its definitional role as a necessary counterpart to human uniqueness, to humanity's own self-decreed, political and ethical, exceptionality from so-called laws of nature.

To point this out is not, though, to support yet another form of biological or ecological reductionism. It is *not* a call to recognize the sovereignty of nature over all human activities, including ethics and politics. This charge is often leveled at radical ecologists, and it is true that some environmentalists may be guilty of aiding and abetting a scientistic reductionism, that they too may be in the sway of that technological enframing that fosters such assumptions, for example, by trying to reduce human politics and ethics to neo-Malthusian matters of ecological carrying capacity and resource depletion. But this is precisely what *radical* ecology is not. It is a political and ecological critique of sovereignty per se, both natural and political. The breadth and depth of this critique is why radical ecology is potentially *the most radical form of politics,* why it offers the most fundamental challenge to the established order of things.

This, again, is why we can say that radical ecology tries to save politics and ethics (and not just the natural world) to recognize their relative autonomy and their vital importance in constituting a good life for humans within, and not constitutionally positioned as a sovereign power above, a more-than-human world.[4] This being so, the question now becomes one of the ways—and there are many possibilities—in which one might envisage the relative autonomy of nature, politics and ethics, of saving them in such a way that all are released into (in Heidegger's terms) their "essence" (into natural and social histories) free from biopolitics and sovereign power and the technological enframing of the world. "Saving does not only snatch something from a danger. To save really means to set something free into its own essence" (Heidegger in Evernden 1999, 68).

Of course, there are countervailing dangers here from a theoretical perspective. First, Heidegger associates "saving" with "letting be" *(Seinlassen),* a turn of phrase that can be, and often is, further misinterpreted in terms of a rather crass notion of passively "leaving things alone" or of contemplative noninvolvement. But this is not what Heidegger means by letting be, nor does it come close to what is intended here. To let something be is to hold open the possibilities of beings appearing in ways that are significant while not simply conforming to our expectations, desires, or definitions. It is to recognize that a being has such potential significance precisely because it transcends (goes beyond) what we would otherwise make of it. Letting be attends to the openness of the world, it "means letting oneself in on the open realm and its openness which each and every thing-that-is stands into, the openness, as it were, bringing that thing along with it" (Heidegger in Vail 1972, 53). In other words, it is to let things "ek-sist," stand out into the world, such as they are, rather than "in-sist" on making them appear as we would want them to be, to make them subject to our dominion.

There is more than an echo of Murdoch here. To let a tree be is not necessarily to make *no* use of it; it is certainly not to merely sit back and abstractly contemplate it, but to encounter it *as that tree*—as Murdoch would say, as something alien, irreducible to our purposes (or any other ends it might serve as a means to), and independent. When we think of the tree in this way, we are already thinking of it ethically, and in speaking of it in this way before others, we are already speaking of it politically in a way that challenges its technological enframing.

To let be, then, is not necessarily to leave alone but to be in community with. To be in community with is not to rule over, nor is it to be made formally equal/equivalent. To be in community with (or as Heidegger might say, to dwell in the neighborhood of) is not even to imply that one must hold that thing close to oneself or one's heart (that, for example, one must become a tree hugger—literally or metaphorically). It is to strive to keep open the possibility of attending to what that being *is* in its (indefinable) essence and also to recognize an *ability to respond* to that being's existence that can imply an ethical *responsibility.* An ethical responsibility is not absolved by definitional niceties of the sort—it is *only* a tree—intended in the sense that it is nothing more than an example of the kind of being that is ethically inconsequential. For here the anthropological machine is at work again. To save is to let things be in a sense that decides nothing for us ethically; it leaves the responsibility of the decision open with all its inherent ambiguities and questions. In this sense, the outcome of our engagement with the tree might still be that we decide to turn it into firewood (just as the actions of the inhabitants as Lascaux turned their prey into meat), but we know that this is not what the tree is in its essence—and it is not inconsequential that we do this, it is not something we should do lightly or thoughtlessly or just because it is demanded of us. The tree in its essence is not something that can be reduced to a common or abstract currency—so many cubic meters of cellulose, so many dollars and cents. And an economy or a society based on such a reduction understands little or nothing of what a tree is; indeed, such a question is entirely meaningless to it.

A second countervailing danger might be that speaking of the "essences" of beings, or of ethics or politics, may be taken as a metaphysical (essentialist) attempt to reconstitute the anthropological machine—for example, to view humanity as something essentially set apart from nature, to think of natural and social history as entirely disconnected processes. (However, as we have seen, when Heidegger speaks of essence here, he is not referring to some [strongly Platonic] metaphysical ideal but to an irreducible reality that we must recognize exceeds and endures beyond any typological attempt to define it.) But this view would be doubly mistaken, since social and natural histories are not processes, nor are they disconnected. Natural and social histories are not processes, at least insofar as this notion might be understood in terms that Benjamin (1973) critiques as proposing either a kind of teleology or historicism (see chapter 1 and Smith 2005a). Such historicism is itself often aligned with a technological enframing and with the exercise of sovereign power (insofar as both natural and social histories are reduced to totalized stories that operate as means of justifying established powers). Natural and social histories are not disconnected, since they are both modes of being-in-the-world that together compose that world. The key point is that the modes of being of politics and ethics and of nature are quite different from each other; neither is *essentially reducible* to the other. As Sandilands (1999, 206)[5] suggests, "Neither political nor animal contain the irreducible truth of the other." More specifically, a politicoethical mode of being is one closely associated with (and most fully developed within) certain human "forms of life" (in a Wittgensteinian, not a purely biological, sense). Humans are, as Aristotle argued and Arendt and Agamben agree, *bios politikos,* beings with an ability to constitute themselves as (differently) human within and through their political forms of life.[6]

This might seem a somewhat unexpected position for radical ecology to take, since it is more usually associated with attempts to *reduce* the perceived differences between human and nonhuman beings, with more biocentric rather than anthropocentric approaches, and with an emphasis on the myriad ways in which human sociopolitical existence cannot be separated from our relations with the natural world. It might also (though quite wrongly) be taken as suggesting that Agamben is actually right to overlook the implications of such self-constituting political activities for nonhuman nature. But what is being argued here is that, ironically, despite the ultimate dependence of political possibilities on nonhuman nature, the only possibility of saving the natural world from the consequences of our increasingly biopolitical form of life and from the epistemic sovereignty of a technological *Gestell* applied to life itself comes from constituting a new politics, from revitalizing politicoethical understandings of the possible relations among humanity and the nonhuman world. These possibilities will not exist if we insist on obscuring the relative autonomy of natural and social histories, if we reduce one to its other. Instead, we need to be mindful of their co-constitutive (although often agonistic) world-forming involvements.

What is more, since nature is not essentially (in and of itself) political, saving nature only becomes possible through rearticulating the relative autonomy of both nature and politics in terms of an ecological ethics, that is, as politically expressed ethical concerns for nature that offer critiques of claims to human sovereignty. This possibility is precisely what forms the core of any radical ecology and precisely what is explicitly missing in Agamben's analysis. Rarely, if anywhere in his work, does Agamben show the slightest concern about the repercussions for animals or their relegation to what has been a permanent state of exception, still less for nature in any wider sense. The possibility remains, though, of a radical ecological take on aspects of Agamben's work, one that develops his analysis of sovereignty and bare life in terms of a broader understanding of saving the natural world, such as that suggested by the work of Evernden. This rearticulation must obviously pay especial attention to the ethical ramifications of the different ways in which such theorists understand the differentiation of human being *(Dasein)* from the nonhuman world.

The Natural Alien: Evernden, Agamben, and Suspended Animation

Evernden's text *The Natural Alien* is arguably the most important philosophical text to begin to address our contemporary human condition from a radical ecological point of view. As already noted, it emerges from a phenomenological and Heideggerian understanding of human being-in-the-world, although it also draws on a much wider range of materials. It is not surprising, then, that Evernden too regards the root of our ecological problems as being intimately connected with the treatment of the natural world as standing reserve and with the emotional and theoretical separation of human subject and natural object that precedes and facilitates this tendency. In effect, he presents us with a philosophical and ecological account of the emergence of the technological enframing of the world and its alienating effects.

This innovative account revolves around an understanding of humans as "natural aliens" destined through their evolving technological abilities to become the sociocultural equivalent of a biologically exotic species. Like Alice in Wonderland, and like introduced species everywhere, humanity finds itself estranged from a world with which it increasingly lacks a shared coevolutionary history. Why? Because, Evernden (1999, 109–10) argues, the pervasive technological innovations that so radically and continually alter our relations to the world mean that "humans cannot evolve *with* an ecosystem anywhere," since, in effect, we are constantly mutating into a new creature. "Figuratively speaking, just as the environment does not know how to cope with the new creature, neither does the exotic know what it ought to do. In other words, the exotic is a problem because it does not know how to comply. It has no sense of context, no relatedness to the community of which it is a part. The creature is suspended in [ecological] ignorance, capable of material existence but not of community commitment."

Humans, like exotics and unlike indigenous species, are ecologically *placeless*. The abilities that make humans so special are inherently ecologically alienating. "Technology displaces its creator and sets him adrift in a world suddenly devoid of sense" (110). This inability to make

sense of the natural world due to our alienation is, Evernden provoca-
tively suggests, what environmentalists are "protesting, not the stripping of natural resources but the stripping of earthly meaning" (124).

Evernden, unlike Agamben, is ethically concerned with both sides
of this equation, with human alienation and with its ecological effects
on the nonhuman world. He also recognizes that we have some leeway
to change or ameliorate this situation. We do not need to "be fatalistic
about our situation" (xii), for our action or inaction does matter. It is
politically possible to counter the notion of human sovereignty over
nature to "abandon the entropic project of planetary domination" (154).
Indeed, *The Natural Alien* explicitly argues that this political possibility is founded on each of us being a "naïve ontologist" (143) who has
a responsibility to dissent when the reality of our experiences of "irre-
pressible wonder at the existence of life" fail to fit institutional realities,
when "the public 'story' becomes one we cannot live through" (143).

That being said, there are worries about the degree to which
Evernden's account of alienation from nature may be too ecologically
reductive. He may claim to be speaking figuratively, but in order for
his analysis of our contemporary human condition to make sense, his
references to humanity's ecological suspension must also be taken literally. At one point, Evernden even compares humans to locusts, another
species that manifests itself in cataclysmic ecological events. In fact,
Evernden claims that we are even more ecologically destructive than
locusts, since "we seem to have maintained our plague phase so long
that we threaten permanent destruction rather than a successional
setback. That is, we now appear as agents of entropy rather than as
inadvertent champions of heterogeneity" (129). Unlike locusts, human
activities do not even have the (supposed) long-term ecological benefits
of maintaining ecological mosaics: we just eradicate everything in our
path with a deadening finality. "We are the global locust, however im-
perfectly we play our role" (129).

Such statements are something of a gift for any critic of radical
ecology who wants to paint it as a misanthropic form of biological
reductionism, although Evernden's claims should obviously be read in
the broader context of his far-from-reductive philosophical analysis,
including his use of Heidegger's notion of technological enframing and
of the political possibilities mentioned earlier. A more comprehensive
response, though, would be to try to pay ethical and political attention
to the nature of this ecological suspension rather than regarding it as

just an unfortunate loss of a previously intimate involvement in natural places. From this perspective, a problem with Evernden's account is that it sometimes tends toward a very unethical celebration of the sovereignty of nature, a call to "acknowledge the strange superiority which our placeful companions enjoy" (154) precisely because they are not ecologically suspended, because their lives are governed by their being caught up in, limited by, and subject to evolutionary and ecological processes and powers. The best humans might do, from this perspective, is to "aspire to some cultural imitation of a life of 'embodied limits'" (154).

This is not an adequate response to our current human condition in terms of either understanding the nature of humanity's ecological alienation or recognizing the political possibilities for saving the natural world. Both our human-centered concerns and any possibility of saving the (natural) world depend on resisting any reduction of humanity to bare life. Indeed, human beings' ecological suspension might be understood as precisely that event which marks the possibility of a nonbiologically reductive ethics and politics. It is only because (at least some) human beings can harbor these (relatively autonomous) ethical and political possibilities that they are at all capable of concerning themselves with saving the natural world. Those of us capable of relating ethically to the natural world are not like locusts, because locusts presumably do not care at all about the damage they cause, nor could they *choose to act* in any way that might reduce this damage. Evernden's almost wistful evocation of a life lived entirely within natural limits unfortunately fails to give enough attention to the dangerous biopolitical consequences of human life without ethics and politics. Ironically, given his total disinterest in ecological concerns, Agamben does offer us a Heideggerian reformulation of this ecological suspension that might, despite its anthropocentric intentions, be adapted to radical ecological purposes.

The Open: Ecological and Human

To recall: Radical ecology is an ethically motivated (that is, noninstrumental) political concern with saving the (natural) world. This saving might be understood as releasing nature from political claims of human sovereignty into the flows of natural (evolutionary and ecological) history, though it does not thereby set humanity apart from nature or decry involvement in nature. After all, humans ek-sist (stand out) in(to) the natural world as beings-in-the-world. Our involvement should, however, be one guided by the ethics of a worldly *phronesis,* not socioeconomic realpolitik. To accomplish this saving, radical ecology must recognize the relative autonomy of ethics and politics and struggle to save these too from their reduction to a biopolitics that is ultimately based in the technological enframing of the (natural) world, that is, the reduction of every aspect of life, human and nonhuman, to standing reserve. The consequences of this *Gestell,* and of biopolitics, for humanity is the stripping of ethical and political possibilities, our reduction to what Agamben calls bare life, a human state of political exception. What we must recognize in our current ecological crisis is that saving nature also depends on generating the possibilities for concerned involvement in the world, which only ethics and politics can offer.

Ironically, then, there would be no ethical concern for the natural world, nor any political possibility of saving the world, if humans were not natural aliens, if we were not, at least in one sense, "ecologically suspended." In *The Open: Man and Animal* (2004), Agamben describes how this suspension might be understood from an admittedly anthropocentric, Heideggerian perspective. Heidegger claims that humans have a capacity to "suspend" themselves from what he terms the disinhibiting ring *(Enthemmungsring),* the enfolding aspects of the world that operate as evolutionarily determined "carriers of significance" for particular nonhuman creatures. These carriers of significance—for example, the right mammalian body temperature that, once sensed by the tick, sets in motion its blood-sucking activities—*captivate* the animal by engaging its specific capabilities. It is in this sense that Heidegger claims that animals are "world poor," unable to free themselves from their spellbound, instinctual attachments to specific aspects of their environments. "Being ceaselessly driven the animal finds itself suspended, as it were, between itself and its environment, even though neither the one nor the other is experienced *as* a being" (Heidegger in Agamben 2004, 54).

Evernden (1999, 168n25) would, quite rightly, be skeptical of Heidegger's use of the term *instinctual* to cover a multitude of potential relations that are by no means automatic or mechanical responses. And one of the reasons to be suspicious of this move is precisely its anthropological intent, its suggestion that (wild) nature has no creativity, no natality, no possibility of initiating something new, or behaving different(ial)ly. However, it is still the loss of this situation, of this suspension of animal being *within ecology,* rather than the suspension of human being *from ecology,* if Heidegger is right, this close, inescapable, involvement within the natural world, that Evernden, like many radical ecologists, regards as a matter of regret. This ecological suspension (originary alienation) may indeed be something to regret insofar as nonhuman life seems to experience a "wealth of being-open, of which human life may know nothing at all" (Heidegger in Agamben 2004, 60), that is, a kind of passionate, naïve, unmediated involvement denied to us—though, here again, the question of whether the loss of our potential for ecological involvement is actually final or complete is very doubtful (see, for example, Abram 1996).

Yet, Heidegger argues, it is this same self-suspension that allows us to see other beings *as* beings and thus to potentially "bring the living thing into the free in such a way as to let the thing which excites 'be'" (Heidegger in Agamben 2004, 58). In other words, the possibility of seeing other beings as things in themselves, beings that might resist their appropriation as mere means for our engaged capabilities, "beings which refuse themselves in their totality" (67), that is, that language too cannot fully capture, depends on this suspension. This same ecological suspension from the environment's disinhibiting ring also marks the self-realization of our own existence as one offering human possibilities and responsibilities. Ethics and politics, the possibility of recognizing (though never fully comprehending) another being as such *and* our potential to free ourselves from captivation by our phenomenal world, to act in word and deed (as Arendt puts it), in ways that are ethically and politically world forming require this kind of suspension, this holding open, these particular avenues of negativity with no use.

Heidegger usually treats this suspension as marking an absolute distinction between human and animal, as in most respects does Agamben, which seems strange, since the point of *The Open* is to critique the anthropological machine. It is true that in his commentary on Heidegger, Agamben (2004, 70) recognizes that under "certain circumstances . . . the animal can suspend its immediate relationship with its environment, without, however, either ceasing to be an animal or becoming human," but this suspension is not one that opens up ethicopolitical possibilities for animals. "The animal is,'" says Agamben, "constituted in such a way that something like a pure possibility can never become manifest within it" (68). Agamben elsewhere defines being-human as *"the simple fact of one's own existence as possibility or potentiality"* (Agamben 2001, 42).

The separation between human and animal, between politics and ecology, is a political necessity for Agamben. This separation must be realized politically if we want to avoid the various forms of biopolitics that constantly threaten to reappear as and when the anthropological machine is redeployed by sovereign power to define a state of exception, to reduce some portion of humanity, and perhaps eventually all humanity, to bare life. What is at stake in the workings of the anthropological machine is the question of "the production and definition of this [human] nature" (Agamben 2004, 22). But what Agamben seeks is a politics that has overcome any need to ground itself in human *nature* precisely because any such politics always founds its claims to sovereign power in the decision (the divide and rule) of who is properly human.

We need to be clear where Agamben's argument leads. He is not, as we might initially think, critical of humanism *because* it constantly tries (and fails) to distinguish the human and the political from animality.[7] On the contrary, what he thinks needs to be overcome is a view of humanity that still conceives of itself and of politics as involving relations that are dependent on our animality *in any kind of terms at all.* What he appreciates in Heidegger is that Heidegger points the way toward a *successful* philosophical and political overcoming of our animality.[8] From Agamben's perspective, we need to rethink the uniquely human political form of life *(bios politikos)* as something that opens pure possibilities of which *zoē* (animality) knows nothing at all.

It is tempting, given the Hegelian and eschatological motifs at the beginning and end of *The Open*—his account of the theriomorphous figures (those human bodies with animal heads) pictured at the end of days (history) in certain ancient religious manuscripts—to view Agamben's own theory in terms of a quasi-Hegelian teleology whereby human spirit *(animus)* is envisaged as the motor of the sociohistorical movement of an anthropological machine that eventually overcomes and transcends *(aufheben)* its own origins, a movement ending in a self-understanding of its political possibilities as an inspired realm no longer dependent on animal nature. In this sense, he would present us with a hyperhumanist (animated/enspirited) materialism that is supposed to inform our political self-understanding. Whether this successfully overcomes the anthropological machine or is just its latest and one of its most extreme iterations, a kind of postmodern political gnosticism, remains a moot point. Why should we not regard Agamben himself as just propounding yet another version of the anthropological machine, one based in the existential possibilities, the openness, of human politics set over and against the animal's instinctive environmental captivation? Many radical ecologists might think that Agamben does not provide a sufficiently coherent answer to this question. This, fortunately, is not quite the whole story.

Saving the (Natural) World?

Agamben's political posthumanism seems very far from Evernden's outlook or that of any radical ecology, but the parallels here are important when it comes to addressing what it might mean to save the (natural) world. Agamben suggests that one of the immediate consequences of humanity being able to free itself from its environmental captivation, that is, of humanity's specifically protopolitical form of ecological suspension, is the ability, in Heidegger's terms, to "let things be." It is only because we are able to suspend ourselves from our environmental disinhibitors that we can think the possibility of letting things be such as they are in themselves. Only thus can we recognize that there is always more to other things than would be revealed as in any way significant if we were always captivated by our naturalistic concerns. Herein, at least for radical ecologists, lies the possibility of an environmental ethics, a letting be that ceases to regard nature as means—even for political ends.

Many radical ecological readings of Heidegger take up this idea of letting nature be (Foltz 1995; Grange in Evernden 1999, 69). However, radical ecologists have tended to employ Heidegger against himself, stressing his later work and drawing out ecological aspects of his notion of "dwelling" in particular. In this sense, they emphasize the incarnality and "naturalness" of human being-in-the-world as the outcome of his rejection of all dualistic metaphysics (Schalow 2006). Humanity, says Michael Zimmerman (1995, 264) "needs a new self-understanding that will eliminate humanity–nature dualism as well as the kind of anthropocentrism that justifies the heedless exploitation of nature. We must learn what it means to let things—human and nonhuman—be."[9] Radical ecology has also tended to emphasize the end of human history

in terms of our self-caused extinction rather than through the triumph
of biopolitics, but there is still clearly a shared interest here in overcoming dualisms and any technological enframing of the world.

A key question is whether the desire to overcome human–nature
dualism that is so central a concern for radical ecology is actually
so very different from Agamben's desire to overcome the anthropological machine that articulates just such dualisms. The answer is that
while each is motivated by very different concerns, they might still be
mutually informative precisely because neither wants to envisage this
overcoming in terms of a simplistic reversion of humanity to animality.
Radical ecology seems to offer the prospect of a new ecologically attuned human life but, if truth be told, rarely pays sufficient attention to
either the ecological suspension necessary to enable politics and ethics
to emerge or to what would happen to these particular human possibilities if and when such an attunement were reached. As Evernden's
work illustrates, and despite its best intentions, radical ecology's tentative articulation of an ecological politics constantly risks falling back
into a naturalistic reduction of politics to some form of captivation, of
"following nature," of natural limits, of becoming subject to the sovereignty of nature in one sense or another.

Agamben's position, on the other hand, claims to offer a life of absolute political emancipation, of human freedom but entirely without an
ecological (or sociohistorical) context, one wherein ecology is somehow
implausibly left behind, abandoned to its own unknowable purposes.
Here politics risks losing any meaningful connection with the natural
world because this supposedly radical position lacks any ecological articulation of politics whatsoever. In Calarco's (2007, 164–65) words:
"Where one might expect a radically post-humanist thinker such as
Agamben to challenge the oppositional and reductionistic determina-
tions of animal life characteristic of Western metaphysics, he has . . .
remained largely content to occupy the human side of the human/
animal binary in order to complicate and rethink the political consequences of essentialist definitions of the human."

What then actually happens to the natural world in Agamben's
"coming community"? He speaks of this future prospect in terms of a
"natural life that is unsavable and that has been abandoned by every
spiritual element—and yet because of the 'great ignorance' [animals'
unawareness of the very possibility of desiring anything other than
their natural possibilities, that is, their environmental captivation] is

nonetheless perfectly blessed." Human and animal are each released
into "their own truer nature," nature apparently being left serene in
its "non-knowledge." The animal will be "let be *outside of being*,"
outside, that is, of the human phenomenal world (Agamben 2004,
90–91).[10] On one level this sounds, as might befit Agamben's spiritual
hyperhumanism, suspiciously like the rapture preached by Christian
fundamentalists, which also envisages the leaving behind (abandonment) of the natural world to save the truly human soul. Agamben,
however, is not envisaging this in terms of humanity leaving the world
at all but of humanity leaving behind its political concerns with its
own nature in order to inhabit a world of possibilities that are not
governed by the workings of the anthropological machine. This, he
thinks, necessarily involves both the letting be of nature as such and
the recognition that there is more to life than the (natural) world as
it appears to human concerns. "How the world is—this is outside the
world" (2000, 105).

How might this inform radical ecology's interest in saving the (natural) world? A clue emerges in the way Agamben (2004, 82) speaks of
this abandonment in terms of Benjamin's notion of the "saved night,"
that is, the "nature that has been given back to itself," to its own transient appearances and rhythms. This giving back to itself first involves
the recognition that there is so much more to nature, to the operations of those concealed rhythms, to life, than how it appears to us—
especially in the very limited mode of appearance forced on nature
in its technological enframement. It also recognizes that we have to
abandon the attempt to represent nature fully or fix it in its relation to
us as having a certain identity. In Heidegger's terminology, the natural
world in itself is neither ready-to-hand nor present-at-hand: it cannot
be fully captured instrumentally or conceptually such as it is.[11] But we
can only come to think this possibility through our ecological suspension. In one sense, and despite Agamben's anthropocentric intentions,
this offers possibilities for truly ethical recognition of the importance
of letting nature be, not just, as Evernden suggests, in terms of saving
those aspects of a world that have meaning for us, but going beyond
this, of recognizing in Murdoch's and Levinas's sense a relation of infinity rather than totality (see chapter 2). We cannot save the world by
bewailing the loss of just those elements that have meaning for us (as
Evernden sometimes seems to suggest); we must recognize that how
the world is, is also *outside* the human world. As naïve ontologists, we

had always felt something of this world that is "alive to us" but had not yet been able to conceive of its implications in ethical terms.

So while Agamben lacks an ecological ethics or any explicit concern for the natural world, while he seems to think it possible that we can inhabit a world where ecology has no political meaning whatsoever, he still points a way to understanding how human and animal, politics and nature, history and natural history might eventually be reconciled. This reconciliation does not entail, as critics of radical ecology (and fundamentalist primitivists) claim, an impossible return to a mythic state of nature or any reversion to animality. It requires that we reject the claims of sovereignty in all its forms, natural and political. It requires the political mastery of politics as an ethically informed practice by all the world's people rather than the (bio)political mastery of the world, the ethical recognition of the necessary openness of politics and nature. This means that those capable of doing so accept responsibility for the (pure) means of political "production" in such a way that they let nature be, free from any claims to sovereign power over it.

Agamben's position is not as Hegelian as it initially seemed because he, like Bataille, concerns himself with what exceeds and resists the claims of any purportedly universal or totalizing dialectic:

> What does this "mastery of the relation between nature and humanity" mean? That neither must man master nature nor nature man. Nor must both be surpassed by a third term that would represent their synthesis. Rather, according to the Benjaminian model of a "dialectic at a standstill," what is decisive here is only the "between," the interval or, we might say, the play between the two terms, their immediate constellation in a non-coincidence. The anthropological machine no longer articulates nature and man in order to produce the human through the [political] suspension and capture of the inhuman. The machine is, so to speak, stopped; it is "at a standstill." And, in the reciprocal suspension of the two terms, something for which we perhaps have no name and which is neither animal nor man settles in between nature and humanity and holds itself in the mastered relation, in the saved night. (2004, 83)

Perhaps we might find a name for this "reciprocal suspension," the ecological suspension that enables ethics and politics and the political suspension that frees nature into the creative exuberance of the saved night. We might refer to this reconciliation too as a form of "suspended animation," not in the sense of bare life nor of a cryogenic stopping of life's rhythms or even of leaving nature hanging over the abyss of ecological destruction, but as an *image* of the ethical and political holding open of life's possibilities for both human and more-than-human worlds, as the ethicopolitical suspension of that originary ecological suspension—those events that initially open the world from its captivation. In this way we might still, naively, attempt to save the (natural) world.

Suspended Animation and the Political Articulation of (State) Sovereignty

How might the ethicopolitical possibilities that emerge from the reciprocal suspension of ecology and politics be realized and sustained? Only, it seems, through a fundamental critique of the *original principle* of dominion and stewardship, that is, of political sovereignty, whether the sovereignty of nature over politics, the sovereign individual, or the sovereignty of politics over nature—that same sovereignty which, in modernity, becomes explicitly associated with the state form.[12] The origins of the modern state's claims to political sovereignty are, as already indicated, philosophically constituted through the myth of a productive disjunction between (a properly human) civil society and the state of nature. This myth deploys partially secularized theological concepts that still, in almost every way, emphasize the exemplary (god-like) status of Man (in the strong metaphysical sense) as a species essentially set apart from and above nature and of men (also understood metaphysically) as "more or less successful repetition[s] of the same" (Arendt 2005, 95). All agree to contractually bind themselves to—and thereby authorize—state sovereignty.

Contemporary iterations of this still evolving political form simply assume, as part of the state's very definition, its original and rightful territorial dominion over nature—today primarily reenvisaged and accounted for in terms of natural and national resources, that is, ultimately as standing reserve. Although the territorial dimensions of these "imagined communities" (as Anderson [1991, 6] defines nation-states) are configured within pervasive discursive regimes that also mobilize "patriotic" images of "fatherland," "motherland," "homeland," and so on, that is, within regimes dependent on the uncritical deployment of mythic notions of the original, inalienable, simultaneous "birth" (*nascence*) of people and (nation) state together. This nationalistic imaginary *suits* the state.

But even though the timeless mythic principles on which state authority is constituted are supposed (de facto) to command universal assent, *everyone* actually knows they are neither timeless nor universal. As Pateman (1985, 168) argues: "Liberal democratic theorists treat the state as if it were a natural feature of the world," but without the hypothetical voluntarism assumed by the original (mythic) social contract, "the emperor is indeed naked."[13] The state form has a relatively recent history. There never was a state of nature or a social contract. State boundaries are inventions. The right of humans to rule the natural world is as politically arbitrary as the feudal notion of the divine right of kings. Of course, it still takes a certain kind of political naivety to state the obvious. But radical ecology constitutes a fundamental political challenge precisely because it refuses to accept the *reality* of any aspect of this *myth* of state sovereignty, whether in terms of sovereignty over human political possibilities or natality of the natural world or national territories. Instead it advocates ethical, nonauthoritarian, nonterritorially delimited relations to the more-than-human world, that is, to adapt Levinas's term, it envisages ecological ethics as anarchic "first philosophy"—a philosophy that can persuasively inform (rather than compel assent to) diverse forms of ecological politics.

Here, once again, Agamben's work offers important insights not only in terms of his critique of the anthropological machine and of the biopolitical reduction of human individuals to bare life but also in terms of the ecological potential of his critique of sovereignty—a potential that exists despite Agamben's understanding politics entirely in terms of community with other humans (see chapter 3) and never considers the possibility of a "coming ecological community" (Agamben 2001). Still, as already indicated, Agamben's work constitutes a fundamental ethicopolitical critique of the very idea of sovereignty, one now finding echoes well beyond Italian radical circles (Virno and Hardt 1996) through writers like Judith Butler (2004) and Slavoj Žižek (2002), albeit one that is absent from environmental discussions of state sovereignty (for example, Litfin 1998; Eckersely 2004; Barry and Eckersley 2005). This absence is not unrelated to the radical nature of Agamben's critique, which would certainly undermine any attempt to recuperate a role for state sovereignty for ecological purposes (see chapter 7).

In several books, most especially *Homo sacer* (1998) and its sequel *State of Exception* (2005), Agamben combines his critical appropriation of Foucault's concept of biopolitics with Schmitt's (1985) account of sovereignty. In this way, he seeks to show how contemporary claims of state sovereignty are complicit in the biopolitical reduction of the sphere of human politics to the technical administration and management of populations. And while Agamben's appropriation of Foucault's notion of biopower is certainly contentious (as several recent essays indicate; see Calarco and DeCaroli 2007),[14] his interpretation of Schmitt plays the key role in his political analysis.

Schmitt's *Political Theology* (2005, 5) opens with his famous definition: "Sovereign is he who decides on the exception"; that is to say, it is the ultimate mark of sovereign power to be able to suspend the normal rule of law and the political order by declaring a state of emergency (exception). Further, since such a suspension is paradigmatically only envisaged under exceptional circumstances (at times of political crisis), the precise conditions of its imposition cannot be predetermined (and hence codified in law or a procedural politics) but depend on an *extralegal/procedural decision* made by the very power that thereby awards itself a monopoly on political power/action. The rule (of law) as an expression of sovereign power declares a state of emergency where "suspending itself, gives rise to the exception and [simultaneously] maintains itself in relation to the exception" (Agamben 1998, 18).

Agamben, like Schmitt, emphasizes how the possibility of this ultimately arbitrary decisionistic assumption of absolute territorial authority underlies all claims of state sovereignty, no matter what kind of political constitution such states espouse. Paradoxically, then, the (state of) exception is precisely that situation that (ap)proves the sovereign power's rule. "What the 'ark' of power contains at its center is the state of exception—but this is essentially an empty space" (Agamben 2005, 86). The declaration of a state of emergency is both the ultimate political act and simultaneously the abrogation of politics per se. Here, participation in the political realm, which from Arendt's (1958, 198) and Agamben's perspectives "rises directly out of acting together, the 'sharing of words and deeds,'" is denied by a political decision to some or all of the population of a sovereign territory, thereby reducing them to a condition of bare life.

Agamben thus reaffirms the Aristotelian description of humans as *bios politikos,* as the kind of beings whose *form of life* is such as to enable (but not compel) them to participate in a political community (and, as Arendt argues, to appear before others as particular persons through that involvement). This possibility is denied in the reduction of human beings to the inhuman(e) condition of bare life, the most appalling example of which, Agamben claims, is found in the concentration camp. Here, the political exception took on a literal and localizable form as a real space containing those whom sovereign power had decided to exclude from the political community (those reduced to bare life) under the auspices of a state of emergency. "Inasmuch as its inhabitants have been stripped of every political status and reduced completely to naked life [bare life], the camp is also the most biopolitical space that has ever been realized" (Agamben 2000, 40). All political and ethical norms were suspended, with the most horrific consequences, since once the camp's inmates were legalistically defined as nonpersons, stripped of their citizenship and any ethicopolitical standing in the eyes of the state, "no act committed against them could appear any longer as a crime" (Agamben 1998, 171).[15]

Since Agamben's analysis is intended to apply to the notion of sovereignty as such, and not just the singular state of emergency in Nazi Germany, this also means that despite its extremity, the camp is far from being an isolated instance. Agamben (1998, 166) regards "the camp as the nomos of the modern," an exemplary form in the negative sense that it was "merely the place in which the most absolute *conditio inhumana* that has ever existed on earth was realized" (166). The specter of the camp reappears wherever sovereign power institutes a state of exception that reduces people to bare life and especially when this state of exception is given a fixed spatial arrangement. Agamben (2005, 3–4) argues that Guantánamo Bay, for example, could only really be understood as a camp (see also Ek 2006; Gregory 2006), an exceptional space for containing detainees denied any recourse to normal legal or political process. Here again, sovereign power is demonstrated (made monstrously obvious) through an "inclusive exclusion," that is, their exclusion (being held in suspension) from the political community is the very mark of their *subjection* to that sovereign power. (As always, Agamben's political purpose here is not to compare the relative degrees of suffering such circumstances cause, since this varies radically from case to case, but to expose their underlying unity of form in terms of their relation to the exceptional and absolute claims of sovereign power.)

The new global (and unending) war against terror used to justify Guantánamo is also indicative of the ways in which what is initially justified as a state of exception, an emergency measure, can easily become the (a)political norm. As Walter Benjamin (2006, 392) remarked, the "tradition of the oppressed teaches us that the 'state of emergency' in which we live is not the exception but the rule." And this occurs precisely where the "political system of the modern nation state . . . enters into a lasting crisis, and the state decides to assume directly the care of the nation's biological life as one of its proper tasks" (Agamben 1998, 174–75). As the state of emergency (declared on the basis of a perceived threat to the state's continued existence) becomes permanent, so the defense of sovereign power in the name of survival becomes its own justification. The political relations *(bios politikos)* on which the state's existence, as a supposedly "natural" expression of a *political* community, were premised are suppressed. Instead, the state deploys its (extra)constitutional sovereign powers to control all serious countervailing political expression. It reconstitutes itself on the basis of the biopolitical management of populations where the diffuse (largely nonlocalized) treatment of the nation's populace as bare life—for example, as so much biometric and genetic information—becomes normalized. In Žižek's (2002, 100) words, we come to inhabit a new world order where the "very democratic public space is a mask concealing the fact that, ultimately, we are all *Homo sacer,*" that is, a world dominated by a hegemonic "postpolitics," the fundamental feature of which "is the reduction of politics to 'biopolitics' in the precise sense of administering and regulating 'mere life.'" This shift only emphasizes that for Agamben, sovereign power is never a creative (constituting) political power (as Schmitt portrays it) but only a (constituted/constitutional) power based ultimately in the ability to suspend, to place in abeyance, ethics and politics as such.

Now, Agamben rarely, if ever, links the biopolitical management of human populations to a critical analysis of the sovereign authority by which states lay claim to nonhuman nature. By tying the state of exception solely to the reduction of politics to bare life, sovereignty's defining moment becomes measured entirely in terms of its human repercussions. It is not even clear in Agamben's terms what the defining act of sovereignty with regard to nature would be, since nature (lacking a political dimension of its own) cannot be so reduced. Nonhuman nature is just that which underlies the territory within which states of exception are delimited. This is clearly not sufficient for a radical ecological critique of sovereignty and still begs certain questions even where more traditional political analyses are concerned, since a nation's claim to sovereign power is always already a claim to authority over ecological communities, mountain ranges, minerals, and so on, and not just people. Yet, even bearing this limitation in mind, Agamben's analysis still has important ecological repercussions concerning both the emergence of specifically ecological states of emergency and concerning the limitations that states may impose on ecological politics.

In a technologically enframed (and politically diminished) condition, crises of all kinds are *manufactured* in the dual sense that they are produced, deliberately or as side effects of socioeconomic processes that constantly transform reality, and employed, as Benjamin argues, as fictions (Agamben 2005, 3) to justify political repression. Discussion of whether the ecological reality of any particular situation merits the suspension of politics and ethics is, to some extent, beside the radical ecological political point (such a suspension must always be resisted), although the question of the extent of sovereign power's involvement in manufacturing a crisis situation, including an ecological crisis like global warming, is clearly not. The real concern is that sovereign power (and, remember, Agamben is thinking primarily of state power) has, as part of its self-definition as "sovereign," accrued the sole right to decide this question.

There is thus a real and devastatingly ironic possibility that the idea of an ecological crisis, so long and so vehemently denied by every state, will now find itself recuperated by the very powers responsible for bringing that crisis about, as the latest and most comprehensive justification for a political state of emergency, a condition that serves to insulate those powers against all political and ethical critique.

We may find that the global war on terror will segue seamlessly into the crisis of global warming, a condition produced by previous technological interventions in the natural world, interventions of a kind that were initially deemed politically unchallengeable by everyone *except* radical ecologists. The growing (political and ecological) danger is that this emergency is used to legitimate further technocratic interventions, to further extend the state and corporate management of biological life, including the continuing reduction of humanity to bare life.

We should be clear what is at stake here: nothing less than the ecological future of the natural world and the ethicopolitical future of humanity. The dry bed of the Aral Sea, the burning forests of Southeast Asia, the devastated landscape wrought by the exploitation of the Athabasca oil-tar sands, the industrial-scale slaughter of seal pups on Canada's east coast, and a million other examples all reveal the likely destiny of the natural world without ethicopolitical intervention. As for the reduction of humanity to bare life, this is, as Agamben claims, already well underway. Here too we find states moving toward the biopolitical management of populations; here too the procedures are justified by "exceptional" circumstances that become the new rule(s). A more spatially and temporally localized (and hence more intense) example might be found in the state of emergency declared in New Orleans after hurricane Katrina. What was portrayed as a failure to predict or manage a natural event led to the ethical and political abandonment of a largely African American urban population and the simultaneous imposition of martial law on that same population. The concern, if Agamben is right, is that the disastrous consequences of such instances increase the likelihood of further technological interventions and the call for more rigorous bureaucratic control and police powers on ever-increasing scales. That environmentalists now frequently find themselves labeled as ecoterrorists, as the new enemy within the state, only supports this contention (Vanderheiden 2005; Miller, Rivera, and Yelin 2008).

It seems that despite defining sovereignty in terms of the foreclosure of specifically human political possibilities, Agamben's critique will become increasingly ecologically relevant as environmental crises continue to move center stage. Yet there are also ways in which this understanding of sovereignty might be extended more directly to the more-than-human world precisely because political sovereignty is never just the exercise of sovereignty over politics. If the principle of sovereignty is to decide on the state of exception—to politically suspend the "negativity without a use," the open texture, excess, natality, plurality, and expressions of singularity, individuality, and difference that are constitutive of politics as such—then, as already suggested, such suspensions affect other fields where freedoms are initiated too. Sovereignty can also be instantiated in the political suppression of "art, love, play, etc., etc." and, from a radical ecological perspective, in the stripping of nature's wildness—its biopolitical reduction to matters of technological control and management, its abstract reduction to use and exchange values, its enframing as standing reserve in order to secure the *survival* of an increasingly (anti)social and (a)political system. In other words, the exercise of political sovereignty in the fields of politics and nature is not just analogous but identical insofar as the original (empty) principle of sovereignty is, in all cases, that of a self-awarded exceptional authority inaugurated in the declaration of a state of emergency but increasingly applied universally as the new biopolitical rule.

And this, of course, is where the anthropological machine also comes into play, manufacturing a complicated series of distinctions that both exclude nature from and include it within the political realm, distinctions that set humanity apart from and above the more-than-human world. This is how the state of nature becomes, by definition, the extrapolitical condition that naturalizes political authority and also that first becomes subject to, and abandoned by, politics. The state of nature is not a natural condition (it is not real) but a creation of the sovereign decision to place nature in a state of exception (of an inclusive exclusion), to redefine nature within the scope of the modern political imaginary. This is how sovereignty works, by "deciding" in order to create from nothing the grounds of its own authority. And insofar as the law is that sanctioned by sovereign authority (sovereignty is, after all, the key legal principle) and represents the *normalization* of politics, then Agamben (1998, 26) can argue that the "exception is the originary form of law." "The law has a regulative character and is a 'rule' not because it commands and proscribes, but because it must first of all create the sphere of its own reference in real life and *make that reference regular.*"

In other words, political and legal authority ultimately rest in decisions about how the law *(nomos)* includes and excludes life *(physis).* "This is why," Agamben (1998, 25–26) says,

> sovereignty presents itself in Schmitt in the form of a decision on the exception. Here the decision is not the expression of the will of a subject hierarchically superior to all others, but rather represents the inscription within the body of the *nomos* of the exteriority that animates it and gives it meaning. The sovereign decides not the licit and illicit but the originary inclusion of the living in the sphere of law, or, in the words of Schmitt, "the normal structuring of life relations," which the law needs.

Furthermore,

> Law is made of nothing but what it manages to capture inside itself through the inclusive exclusion of the exception: it nourishes itself on this exception and is a dead letter without it. In this sense, the law truly "has no existence in itself, but rather has its being in the very life of men." The sovereign decision traces and from time to time renews this threshold of indistinction between outside and inside, exclusion and inclusion, *nomos* and *physis*, in which life is originally excepted in law. (1998, 26)

The law feeds off the life of the polis on which it depends for its very existence and the regulation of which is its entire purpose. A radical ecological perspective would point out that law nourishes itself not only on the political life of human beings but on life *(physis)* in the wider senses associated with more-than-human nature. This hierarchic relation means that nature also becomes increasingly subject to regulation and confinement, even in those areas specifically defined as wilderness, like national parks (Hermer 2002).

Nature as such *(physis)* becomes *fugitive:* fleeing the domineering tenets of a law to which it is made subject but in which it has no standing or interests (Stone 1988); exiled from any constitutive role in a political community defined wholly in terms of human citizens (Smith 2005c); sought out by, and taking flight from, attempts to capture and objectify it in totalizing discourses that subsume it under scientific and economic laws, laws that reduce its creative diversity to formulae confirmed by the repeatability of human experimentation (Evernden 1999) or figures calculated to fulfill human desires—reduced to a state of exception not as bare life but as *raw material.* As Agamben (1998, 37) points out,

> the state of nature and the state of exception are nothing but two sides of a single topological process in which what was presupposed as external (the state of nature) now reappears, as in a Möbius strip or Leyden Jar, in the inside (as a state of exception), and the sovereign power is this very impossibility of distinguishing between outside and inside, nature and exception, *physis* and *nomos.*

The myth of sovereignty's origins veils this zone of indistinction, but recent environmental concerns have begun to reveal both its emptiness and its destructive consequences. This, again, begins to explain why radical ecology might claim to be regarded as potentially the most radical form of politics, why it offers the most fundamental challenge to the established order—the political constitution of modernism. Almost all other contemporary forms of politics retain at their heart an explicit or implicit notion of sovereignty that remains fundamentally unchallenged, a residual (originally theocratic) ideology of purportedly justifiably accumulated powers whereby one sociopolitically defined body takes upon itself the right to decide what for others are matters of life and death. Often, even otherwise radical perspectives explicitly accept some form of sovereignty as a political necessity, as something inevitable (due to the movement of history, the laws of social science and/or nature) or at least as politically expedient. More importantly, in the present context, even the most egalitarian humanist political theories still assume a political myth of human sovereignty over the natural world, a self-acquired "right" to define a boundary of political and ethical concern that subsequently treats all outside that boundary as bare life or raw material.

Even many varieties of anarchism that otherwise offer the most thoroughgoing critiques of every hint of sovereign powers have often accepted, indeed celebrated, human domination over the nonhuman world. For example, Kropotkin (1913), who famously argues that "ethical" concerns are not the preserve only of humans, still often praises the suppression of nonhuman nature in terms of the clearance of forests, the drainage of marshes, the highways and railroads that pierce the mountains. These human examples of sovereign power over nature seemingly need no further ethicopolitical justification because the idea/ideology that nature exists solely to serve human needs is all pervasive.

This kind of inclusive exclusion, the relegating of some inhabitants of the world to a state of exception, is, as Schmitt's work suggests, the ultimate (and ultimately unjustifiable in anything but its own terms) foundation of all claims to political sovereignty and, of course, of Agamben's anthropological machine. But claims to sovereignty are not thereby, as Agamben suggests, just antithetical to emancipatory human projects; they are not just articulated as a biopolitics applied only to the human realm. Rather they are, as already argued, inseparably connected to that other biopolitics that has established a state of affairs, a technological *Gestell,* based on ecological transformation and devastation on an unprecedented scale. This is why radical ecology develops an ethicopolitical critique of, and direct challenge to, the most fundamental humanist/modernist political assumption, that of human sovereignty over nature.

The fundamental nature of this critique in no way entails a reactionary form of ecological fundamentalism, a reversion to a supposedly prepolitical natural order ruled over by nature's own sovereign powers. On the contrary, it is radical precisely because it is anarchic in its repudiation of all claims of sovereign power and because it explicitly recognizes that the solution (if there is one) to this anthropogenic ecological crisis lies within ethics and politics. But this has to be a politics

differently conceived, not one of retaining or even extending humanity's sovereign power and mastery over the world but of eliciting the
social and ecological possibilities inherent within political action and
grounded in ethical concern for others (human and nonhuman): in the
saving power that also emerges (like hope from Pandora's box) last of
all from our ecological suspension.

Radical ecology seeks to realize the reciprocal suspension of ecology
and politics; it is the expression of the suspended animation already
described. It also recognizes that what Agamben (2001) so hopefully
describes as *The Coming Community* will never actually become a
worldly possibility unless and until politics is reenvisaged ecologically. This is why it offers such a vital political challenge and why it
is ultimately dependent on the possibility of an environmental ethics,
that is, in our abilities to express concerns for those nonhuman (as
well as human) others whose existence "takes place" all about us. And
again, this cannot be legislated for or guaranteed by sovereign powers, even (especially) in terms of some abstract interspecific egalitarianism (Smith 2001a), but will emerge only through a transformative
politics that recognizes the constitutive (not constituted/constitutional)
world-forming powers of different forms of life, their infinite ecological
potentials. However naively, this requires that we become alive to the
world's possibilities through recognizing that it is, after all, alive to us.

Latour's Political Settlement and Post-Humanist Politics

*What term other than ecology would allow us to welcome nonhumans
into politics.*

—Bruno Latour, *Politics of Nature*

Agamben is not alone in recognizing the *constitutional* role that nature
is forced to play. Neither is he alone in the search for a posthumanist
politics. For example, in his recent *Politics of Nature,* Bruno Latour
(2004) argues that modernity has been characterized by, and dependent on, a conceptual *separation of powers* that serves political ends.
Maintaining the objective neutrality of nature (and Science as *the* privileged form of natural knowledge) is precisely what allows nature to be
deployed to put an end to political debate.[16] This is how claims about
natural rights, natural law, natural orders, and so on, work. Of course,

these accounts of nature are themselves contestable, but the fact that there is a natural order of things regarded as separable from human values has held, and arguably still holds, sway within modernity. In a way, then, Latour (2004, 28) claims, there "has never been any other politics than the politics of nature, and there has never been any other nature than the nature of politics," since nature, qua an external, objective, order, was always a body maintained for political purposes. Interestingly, Arendt too partly prefigures this argument. In a footnote to *The Human Condition,* she remarks that when the Royal Society "was founded, members had to agree to take no part in matters outside the terms of reference given it by the King, especially to take no part in political or religious strife. One is tempted to conclude that the modern scientific ideal of 'objectivity' was born here, which would suggest that its origin is political and not scientific" (Arendt 1958, 271; see also Shapin and Schaffer 1985).

It is this *settlement,* this division of powers that, Latour argues, is now changing through what he considers a crisis of objectivity rather than an ecological crisis per se. For him, the unintended, and as yet largely unrecognized, benefit of an emergent political ecology is precisely that it exposes the supposedly objective political roles that nature has always played. Despite the ways in which the philosophies associated with environmentalism usually remain embedded in this same modern settlement, their practices are, Latour argues, entirely dependent on, and help express, the political mobilization of different (plural) "natures" within hybrid networks. This is how environmental concerns like the depletion of the ozone layer cut through the political paralysis concerning nature. They exemplify the ways in which humans and nonhumans commingle in complex ways within networks that include all kinds of interconnected forces, powers, discourses, institutions, and so on, all of which get caught up in the *action.* Latour offers a very different form of posthumanism from Agamben, one that initially seems much more obviously entangled in the world, less infused with any potential posthumanist gnosticism. Latour uses the term *actants* to refer to these tangled nodes of activity that are neither objects nor subjects as traditionally understood, the diverse and constantly changing members of his "pluriverse," which challenge the absolute distinction between politics and nature.

For Latour (2004, 22), the "risk-free objects, the smooth objects to which we had become accustomed up to now, are giving way to risky attachments, tangled objects"; what used to be regarded as "matters of fact" are becoming "matters of concern." The latter, unlike the former, "have no clear boundaries, no well-defined essences, no sharp separation between their own hard kernel and their environment. It is because of this feature that they take on the aspect of tangled beings, forming rhizomes and networks. In the second place, their producers are no longer invisible, out of sight; they appear in broad daylight, embarrassed, controversial, complicated, implicated, with all their instruments, laboratories, workshops and factories. Scientific, technological, and industrial production has been an integral part of their definition from the beginning" (24). These matters of concern cross over in innumerable ways between what were previously regarded as different universes, especially between nature and politics. "Finally, and this may be the strangest thing of all, they can no longer be detached from the unexpected consequences that they may trigger in the very long run, very far away" (24).

Several themes converge here that are still to be addressed: First is the idea of a posthumanist politics, that is, in Agamben's terms, a politics no longer dependent on the anthropological machine or the sovereign exception of nature, and in Latour's terms, a politics no longer dependent on the modernist settlement between nature and politics, facts and values. Both Agamben and Latour would agree that this requires a politics no longer based on the absolute (metaphysical) separation of humanity from the natural world, although for Agamben a separation still exists in and through politics as such. But how does this compare with Latour's perspective, and what happens to the idea of politics as a "pure means" if it becomes ecologically entangled, if it becomes a political ecology? What form do such entanglements take, and how do they affect the kinds of negativity with no use that compose politics as such? What happens to political action in an Arendtian sense (see chapters 1 and 5) and individual freedom within hybrid networks? Does a posthumanist politics like Latour's elide the difference between politics as such and political systems and processes?

Second, what form(s) might a posthumanist politics actually take? For example, will it be democratic? Anarchic? Will it invoke a new political constitution, or can it be purely constitutive? What changes might be made to the anthropologically exceptional model of a democratic citizenry if both the nature of political action and the political activities of nature are reconsidered and rearticulated? Third is the issue of unpredictability, of unexpected consequences, which for Arendt are an inevitable outcome of political action but for Latour are a consequence of the complex entanglements of actants. How might such unpredictability be linked to an ecological politics? Fourth, what happens to ethicopolitical responsibilities (and indeed to the very possibility of ethics) given such unpredictability and such different (post)humanist understandings of individual actions/networked actants? How are matters of (ethical) *concern* constituted? What advantage might there be for either ecological politics or environmental ethics if, to the extent they gain a foothold in Latour's posthumanist political world, they also seem to have lost both the object of their deliberations (a nonhuman nature that is genuinely "Other") and the (subjective) locus of their concerns?

Each of these themes needs to be considered in relation to emerging ecological communities that would reject the principle of sovereignty as such. The challenge, on the one hand, is to imagine realizable forms of politics that reject human exceptionality (reliance on any strongly metaphysical or supposedly apolitical form of humanism) yet still retain the creative possibilities opened by politics, art, love, play, wildness, etc., etc., and are still informed (but not ruled over) by ethics. On the other hand, this posthumanist politics must avoid dissolving individual ethical and political responsibilities in amorphous hybrid systems or networks that succeed in subverting anthropological distinctions only to expel hard-won freedoms and singular concerns. The danger is that some forms of posthumanism might just employ systems theory and cybernetics to theoretically reduce beings' capacities to initiate change to side effects of ongoing processes, to reduce people to biopolitically managed populations and nature to environmentally managed ecosystems. It is not yet clear to what extent Latour's own politics of nature might fall into a similar trap. Here, though, a comparison between the complex ecologically centered systems theory of Ulrich Beck and the post-Heideggerian political framework created by Arendt might prove informative.

5 RISKS, RESPONSIBILITIES, AND SIDE EFFECTS
Arendt, Beck, and the Politics of Acting into Nature

And *who* are we? . . . the fact that attempts to define the nature
of man lead so easily into an idea which definitely strikes us as
"superhuman" and therefore is identified with the divine may cast
a suspicion upon the very concept of human nature. On the other
hand, the conditions of human existence—life itself, natality and
mortality, worldliness, plurality, and the earth—can never "explain"
what we are or answer the question of who we are for the simple
reason that they never condition us absolutely.
—Hannah Arendt, *The Human Condition*

Who Acts?

"And *who* are we?" Arendt's question is addressed to all those capable
of understanding it but resonates deeply with any politics that refuses
to define a conclusive, once-and-for-all answer that would distinguish
the properly human from the improperly inhuman. Who are we, for
example, who express our concern to save the natural world? And this
is, as Arendt makes clear, a very different question from asking *what*
we are, because this "what" is precisely an attempt to define us as
Homo sapiens, 61.8 percent water by weight, gene machines, Mexican
citizens, Marxists, homeostatic biological systems, unemployed, close
evolutionary relatives of the chimpanzee, and so on. No such list could
ever define who we are—which is not to say such definitions are ir-
relevant but that we are so much more and other than this or that
particular definition allows and that our singularity is not captured
by even the most inclusive taxonomies or extensive lists of predicates.
And this is so of us "whatever being" we are, whether whale, hare, elm,
or bee, although our beings are so very different and open to different

possibilities. Who we are is a question concerning our being in its singularity and in our community with and for others—we—the denizens of the Earth. And for some of us, some of the time, this is also an ethical and political question.

Even (or perhaps especially) our freedoms do not define us—our natality, our lives, our wildness—because, as Arendt argues, our condition (and she, of course, is referring solely to the human condition) is unconditional, which is not an attempt to claim some absolute existential freedom (especially not in a Sartrean humanist form), but only, as she says, in the sense that the circumstances in which we find ourselves never condition us absolutely. In other words, there is a certain undecidability about how we respond to our condition, where "undecidability is not the opposite of a decision, it is the condition of a possibility of a decision" (Derrida in Caputo and Vattimo 2007, 139). And this, for Arendt, is paradigmatically what politics is about: it is the condition of possibility associated with forms of life *(bios politikos)* that human communities *can* aspire to and attain. (This, once again, as Arendt makes explicit, is not to say that politics is a defining feature of being human but a *conditional* possibility.) Politics is the condition of undecidability (of pure means) within which we can make decisions (in the sense of following one path or another)—to speak out or hold our tongue, to demonstrate or acquiesce—decisions that, once made, reveal in public through our words and deeds something of who we are, in our singularity. This is politics *as such,* which should not be confused with any specific political system or with political systems in general. To the extent that politics is systematized, reduced to a process, it ceases to have the character of a condition of possibilities, of initiating freedoms; it ceases to reveal who we are; it just demonstrates what we are for political purposes—a voter, a Party member, a conservative, and so on. Our individual ethical and political responsibilities are always to be more or other than that.

Throughout her life, Hannah Arendt was concerned with tracing the factors that might lead to either the acceptance or abdication of ethical and political responsibility by ordinary people in the often extraordinary circumstances generated within modern societies. This, for example, is how she approached what seemed most difficult to understand about the comprehensive overturning and collapse of values and standards that accompanied and contributed to the Nazi's rise to power—the apparently thoughtless and conscienceless acceptance by so many of rules and regulations that flatly contradicted previous notions of decency and humanity. In her famous account of Eichmann's trial in Jerusalem, she detailed the ways in which even this "architect" of the "final solution" still resisted any attribution of personal responsibility for his central role in the deaths of millions. The chilling "banality" of Eichmann's evil lay in his apparent incapacity to think for himself or even begin to understand what ethical relations to others might involve other than following the norms, rules, and expectations associated with the roles he occupied in that most unethical political system. He thought only in clichés and denied, despite all the evidence to the contrary, that there was ever any undecidability about what he could and did do; he seems to have subsumed "who" he might have been under "what" he thought he was—an efficient administrator, an obedient servant of the state, and so on.

Eichmann's case exemplifies, albeit in an extreme fashion, the appalling consequences of abdicating all personal responsibilities, of a human being who refuses to consider *who* she or he is. This is why Arendt's writings so often focus on the intimate relations among individual responsibility and judgment and social and political circumstances, among thinking, feeling, and acting. However, especially toward the end of her life, she also occasionally addressed environmental issues as newly emerging concerns that directly affected that human condition. Her work has been recognized as potentially informing green politics in several areas (Whiteside 1994; Drucker 1998; Sandilands 1999; Smith 2005b, 2006).[1] In an essay called "Home to Roost," written in the year of her death, she criticized the thoughtless association of ideals of progress with the irresponsible consumer culture that "went on at the expense of the world we live in, and . . . the objects with their built-in obsolescence, which we no longer use but abuse, misuse, and throw away." She also saw the "recent sudden awakening to the threats to our environment" as a "first ray of hope" (Arendt 2003, 262), an indication that some at least might be beginning to think about their wider responsibilities.

For Arendt, the capacity to act is the fundamental feature of political and individual existence, of living among other humans and yet being someone different from all others. Humans may *labor,* expending their biological powers in creating and maintaining life, they may *work* to produce artifacts, but it is only through *acting* and its corollary *speaking* (only in her deeds and words) that the individual reveals herself as *who* she is, as someone with a unique personal identity within a public arena. Acting involves taking an initiative, making a beginning, creating (or at least contributing to) the political and ethical conditions of our existence. Arendt (1958, 198) develops an understanding of the political aspects of the human condition as something that, through action, humans can create among themselves. The "political realm rises directly out of acting together, the 'sharing of words and deeds.'"

It is, of course, the political decision to exclude some from this possibility, to reduce human lives to labor (mere survival) and work, to *what* rather than *who* they are, which exemplifies, in Agamben's terms, the sovereign exception producing "bare life." Labor and work may transform the earth, but it is acting—the "single instances, deeds or events, [that] interrupt the circular movement of daily life" (Arendt 1993, 43)— that, through its political effects, creates a condition that Arendt takes as peculiarly human: "history." But whereas history began, Arendt contends, in Homeric Greece, as a way of ensuring the remembrance of great events, a way of extending the procreative immortality of nature by letting these singular interruptive acts and the individuals performing them live on, it becomes something else in the modern age. History becomes party to "the world-alienation of man" (53), no longer a matter of recording singular events but a single, all-comprehending "man-made" process, an expression of technological society, "the tremendous structure of the human artifice we inhabit today, in whose framework we have even discovered the means of destroying it together with all non-man-made things on earth" (54).

What began with the substitution of "mechanical processes for human activities—laboring and working . . . ended with starting new natural processes" (Arendt 1993, 57). While premodern societies multiplied human labor by harnessing wind and water power and industrialized society put natural forces to work as "man-made means of production" (57) such as steam engines, we have now entered a new stage where "natural forces are let loose, unchained, so to speak, and where the natural processes which take place would never have existed without direct interference of human action" (57). The example Arendt uses is that of nuclear technology, but her point would apply equally to current forms of genetic manipulation, nanotechnology, and so on. We have, Arendt says, "begun to act into nature as we used to act into history," and though we cannot create nature as such insofar as we initiate new natural processes, "we 'make nature' to the extent, that is, that we 'make history'" (57).

This has profound implications for both nature and society: we begin to understand nature itself as a process akin to history that we seek to manage, direct, and control in its entirety. Where industrial society sought to mechanize work in terms of the fabrication of objects, still to some degree regarding nature as the source, the "giver," of raw materials, now humanity comes to regard itself as having nature "in its gift," as belonging to us as a collection of processes we can alter and incite to suit our purposes. In other words, nature becomes envisaged and used as merely a part of an indefinitely malleable human condition. This might seem to be liberating, since making nature part of the human world obliterates "the defensive boundaries between natural elements and the human artifice by which all previous civilizations were hedged in" (Arendt 1993, 60). But this is a very dubious liberty for a number of reasons. Most importantly for Arendt, it marks both the introduction of a new kind of uncertainty into nature, an uncertainty inherent in acting itself, and a novel and more radical form of world alienation. It is also clearly destructive to nature understood as something that can be regarded as more, or other, than a human-controlled process and raises vital concerns regarding responsibility for "acting into nature."

Why is acting into nature necessarily a harbinger of uncertainty? Acting, unlike work, never leaves a completed end product behind it; rather it makes a creative (or sometimes destructive) intervention in chains of events the eventual outcomes of which are entirely unpredictable. Acting is, in it own terms, futile (that is, leaky and lacking permanence), intangible, and fragile, and yet an act's effects multiply beyond all reckoning and in this sense at least achieve a form of endurance "whose force of persistence and continuity in time is far superior to the stable durability of the solid world of things" (Arendt 1958, 232). An action, unlike a particular product, has no end. But this means that by acting into nature, we introduce the same kind of human-induced instabilities, the same "unpredictability into that realm which we used to think of as ruled by inexorable laws." And, as Arendt makes plain, these effects are not simply due to a lack of foresight or prudence but are an ineradicable feature of setting into motion events the ongoing consequences of which there is no possibility of calculating.

Arendt might be said to present a philosophical account of a situation closely related to social theorist Ulrich Beck's influential formulation of *The Risk Society* (1992). Beck too argues that we are entering a new form of society characterized by the unpredictable implications of technological interventions in nature on a global scale. Chernobyl, BSE, global warming, and so on, are not simply accidents; they are not incidental but inherent, though certainly unintended, consequences of the risk-taking on an unprecedented scale that now characterizes modernization itself. They are systemically intensified *repercussions* of wholesale industrialization. These entirely human-created risks that emerge with the initiation and interlinking of new "natural" processes threaten the very existence of human society and "*all* forms of life on this planet. The normative basis of their calculation—the concept of accident and insurance, medical precautions and so on—do not fit the basic dimensions of these modern threats. Atomic plants, for example, are not privately insured or insurable. Atomic accidents are accidents no more (in the limited sense of the word "accident"). They outlast generations" (Beck 1992, 22). They are, one might say, the (technologically induced) exceptions that become the rule.

Beck describes the ways in which the unprecedented scale and synergy of these interventions and their potentially disastrous repercussions erase the boundaries between nature and politics. Environmental risks are, in Beck's terminology, politically *reflexive,* that is, they are forms of self-endangerment through which nature's responsive activities themselves become part of the social and political fabric. Nature is no longer something external but "*because* it is a nature circulating and utilized within the system, nature has become political, even at the hands of objective (natural) scientists" (82). For example, the fallout from Chernobyl was political as well as radioactive, and all subsequent environmental debates have also included concerns about the trustworthiness of scientific risk assessments, the motivations and impartiality of scientists, and so on. Of course, most environmentalists recognize that nature is no longer a world apart, something entirely separable from human influence, but Beck takes this insight further. Today, Beck states, nature is "a highly synthetic product everywhere, an artificial 'nature.' Not a hair or a crumb of it is still 'natural,' if 'natural' means nature being left to itself." Nature has "become a historical product" (81). In Arendt's terms, it has become part of history understood as a process.

While not wanting to overemphasize the similarities between Arendt's and Beck's positions, it is pertinent that in recent interviews Beck has described reflexive modernization in terms of action. "For me, reflex is *action,* action directed backwards; a process of alteration that begins to alter itself, to progressively become a new process. One of the key effects of this is that it introduces turbulence into institutions. This is true whether experts register it or not, but in fact their initial obliviousness often plays an important contributing role" (Beck in Beck and Willms 2004, 33). While reflexivity might incite and include reflection, that is, thinking about what has and could happen (and also *reactions,* which is how Beck interprets environmental movements themselves), such thinking is never in a position to predict the future ramifications of these actions. Attempts to deal with these new dangers through the science of risk analysis misunderstand the nature of the problem, which, contra the dominant perspective, is not amenable to calculation or rationalization. To believe that it is, says Beck, "amounts to pretending that there is no such thing as the unknowable future. It denies in effect that such kinds of risk can exist and only makes them worse" (32). It makes them worse because it gives us the illusion that we know the likely results of our acting into nature while, as Arendt (1993, 60) also argued, "no engineering management of human affairs will ever be able to eliminate [such unpredictability]."

Of course, there are clear differences between Beck's and Arendt's accounts. For a sociologist like Beck, this new "reflexive modernity" with its potentially catastrophic repercussions is a consequence of the globalization of social systems and technologies, their scale, interdependencies, and synergistic relations. For Arendt, it is actually something different about the form of this acting into nature—that humans are now initiating *new* natural "processes"—that makes them so dangerous. Beck's analysis is in many respects more convincing, although this difference is partly a matter of emphasis and both agree on the increasing inseparability of nature and politics, a situation "where man, wherever he goes, encounters only himself" (Arendt 1993, 89). A situation that precisely describes the danger Heidegger saw regarding the technological *Gestell* (see chapter 4). However, the corollary of this "politicization" of nature is explicitly for Beck but only implicitly (if at all) for Arendt, that *politics will, from now on, be inherently, not just accidentally, a politics of nature.*

Action as a Politics of Nature

The idea that all politics might now be regarded as inherently ecological is an extraordinary conclusion, one that most political theorists have still to consider, and also one that needs some qualification. Without doubt, the reason this seems such a bizarre and exaggerated statement is that, as with Arendt's own account, modern understandings of politics regard it only as the field of human action and interaction. The *initium* is always human, and the realm in which their political effects ramify is one of human interrelations, that is, a body politic. Yet, Beck (like Latour) might argue, if nature has become merely another encounter with the human, and if nature's reactions become part of and party to the unpredictable chain of events that follow human interventions, then nature too might be said to have become an integral aspect of that "space of appearances" that, for Arendt (1958, 199), characterizes politics.

This does not necessitate ascribing any kind of conscious agency to nature, but it does mean that nature becomes much more than the passive backdrop against which human politics plays out or the resources that politicians fight over. The body politic also becomes a politics of various human, nonhuman, and hybrid bodies that, as Latour argues, are no longer entirely natural. And if nature (and not just human politics) is now marked by unpredictability rather than characterized by immutable, objective laws, then nature can no longer serve, as it so often has, as a supposedly incontestable limit that defines and constrains political possibilities. For something to be described as natural would no longer mean that it is politically inevitable or irresistible; rather it becomes an opening onto questions of the political desirability of human actions and activities.

Taken in this way, what seems to be (but is actually rather more than) a reflexive extension of Arendt's own analysis appears to offer possibilities for human freedoms akin to those associated with philosophical theses on the death of God and Man—a point endorsed by Latour (2004, 25–26): "When the most frenetic of the ecologists cry out, quaking 'Nature is going to die,' they do not know how right they are. Thank God, Nature is going to die. Yes the great Pan is dead. After the death of God and the death of man, nature, too, had to give up the ghost. It was time: We were about to be unable to engage in politics any more at all."

Such a statement hardly seems ecological! But Latour argues that the rise of political ecology is paradoxically, and quite contrary to expectations, initiating the death of Nature understood as an overarching unified, objective, and socially transcendent order (Nature with a capital N), a death that actually facilitates a condition of political undecidability, that is, politics as such. And this seems plausible if, by Nature, we mean a metaphysical totality understood to be an original source of incontestable (sovereign) laws that are then deployed to serve political ends. The philosophical weakening of such a notion in accordance with Vattimo's ideas (see chapter 1) might be regarded as an important, indeed necessary, step in creating a more secular ecological politics. But it is by no means clear why radical ecologists (or even certain deep ecologists) would quake at such a suggestion if the metaphysical and political limits of such an analysis are understood—because the living, diverse, wild, natural world that environmentalists are interested in saving is not at all the same as the state of nature (Nature), the imaginary, politically created, metaphysical myth that is the main target of Latour's reflexive political ecology.

In other words, Latour's apocalyptic pronouncements about the death of nature should not be read literally (as fundamentalist forms of social constructivism [Blühdorn 2000] might be wont to do). Like Beck (and many social and political theorists, including, for example, Lefebvre [1994], and even many environmentalists like McKibben [1990]),[2] Latour believes that the scale and intensity of human activities and interventions mean that it is no longer justifiable to think of a natural world entirely separable from or uninfluenced by humanity, a recognition that, more than any other, might be taken to signify the threshold of postmodernity (Smith 2001a). But, having said that, even smog-filled skies are still illuminated by the morning sun, acidic rain still falls on trees and streams, and the more-than-human creatures of the world still voice their presence even as species after species is forced to extinction.

One might say that everything is changed by this recognition of the end of Nature, and yet everything that matters, that calls for our attention, remains *as it is* (Smith 1999a, 2001c). Despite Latour's rhetorical flourishes and his works' "provisional appearance of radicality" (2004, 7), for example, his apparently contradictory claims that *"political ecology has nothing to do with nature"* (5), that it has never "had anything to do with nature, with its defence or its protection" (5), and that it must "let go of nature" (chapter 1), these need to be understood in this very specific way: Latour is claiming that a genuinely political ecology cannot use Nature as an incontestable standard or norm according to which, and in order to protect the purity of which, the political realm should be ordered—it cannot *"continue to use nature to abort politics"* (19). In other words, a radical political ecology cannot deploy a politically *sovereign* notion of Nature.[3] In this respect at least, the abandonment of Nature to which Latour refers is not so dissimilar to that proposed by Agamben insofar as he too suggests laying aside the metaphysical claims of an idea(l) of Nature that is politically deployed by the anthropological machine in order to paralyze politics as such. This should, by no means, be anathema to any ecological politics, though it certainly challenges the many attempts made by environmentalists to bypass politics as such, for example, by moving straight from the beliefs of deep ecology and/or the findings of scientific ecology to instituting governmental policy.

And so, three very different perspectives, Arendt's political philosophy, Beck's theory of risk society, and Latour's science studies, all seem to come to similar, if not entirely compatible, conclusions. All make the inherent unpredictability and risks of acting into nature part of their analysis, and all recognize that the consequences of this action have effects on the modern settlement between nature and politics. Having said this, the notion of action associated with Arendt's politics is quite different from that associated with Beck's notion of reflexivity and even further distanced from the hybrid posthumanism of Latour's actants. This increasing distance from Arendt, via Beck, to Latour reflects the progressive dissolution of the nature–politics boundary in each theory—the increasing permeability of interaction envisaged between these (now *dis*established) regions. It also, though, seems to involve an increasingly vague and expansive notion of action to the extent that Latour's definition of an actant is "quite simply, that *they modify other actors through a series of trials that can be listed thanks to some experimental protocol"* (Latour 2004, 75). There seems little space here for the subtlety of Heidegger's (albeit anthropocentric) analysis of the moment of one's suspension from ecology (see chapter 4) or the natality or undecidability that are, for Arendt, constitutive of political action. Unfortunately, questions of ethics and ecological responsibility also seem to be radically altered if not entirely eradicated by such a move.

Who Cares?

Action, for Arendt, is that creative mode of human "being" that constitutes the collective space of appearances through individual performances. Politics is "acting in concert." Its mode of operation is that of power (*not* force) actualized only "where word and deed have not parted company" (Arendt 1958, 200). Acting into nature occurs because modern science becomes a form of organization that, however apolitical it aspires to be, "is always a political institution; [and] where men organize they intend to act and to acquire power." In this case, they "act together and in concert in order to conquer nature" (271). This conquest might be regarded as primarily a political (rather than an epistemic) "success," one that comes at the unfortunate cost of making nature appear as a human-made process, albeit one infected by unpredictability. Understanding our modern human condition in this way means that the environmental repercussions of acting into nature are politicized: one could almost say that today hurricanes (like Katrina) are no longer "acts" of God or Nature but are ineradicably political.

This is a vital though contentious insight because once accepted, as Beck makes explicit, environmental changes can no longer be regarded as externalities to be managed by neutral scientific experts, especially by so-called risk experts who only compound the problem. Such risks are actually consequences of the politics inherent in the organization of science and technology, a politics only now becoming visible as the modern division of powers becomes unsettled and, again like the Emperor's new clothes, increasingly transparent. Environmental issues are then neither accidentally nor incidentally but *inherently* political. They are not just the subject of political debate but are *caused* by politics, by acting into nature as if it were or could become a human-made process. And for Beck, though not for Arendt, nature acts back directly and indirectly, reflectively and materially, into the social world including politics, into the indeterminate future of those spaces of appearance that were themselves once supposed to determine and exemplify what the good life might be.

Beck argues that this now humanized nature, whether conceptualized explicitly as risks, such as those associated with global warming, or operating in as-yet unsuspected ways, initiates a new form of politics: a *reflexive modernity* that dissolves its own taken-for-granted premises. This erodes not only the settlement of powers between nature and politics but also the forms of political organization that previously characterized modernity, including those of class and nation-state. Such "zombie concepts . . . where the idea lives on though the reality to which it corresponds is dead" (Beck in Beck and Willms 2004, 51–52) will, Beck claims, no longer suffice to understand the dynamics of contemporary politics in risk society. For example, nation-states are no longer able to offer their citizens protection from the unpredictable global risks that extend well beyond their territorial borders, risks that synergistically ramify through and among social and environmental systems. And if this is true, of course, much of their philosophical raison d'être also vanishes. Instead, Beck claims, we face a contemporary reality marked by processes of deterritorialized struggles, increasing individualization, and banal cosmopolitanism.

These changes clearly have profound impacts on the apportioning of political and environmental responsibilities. For Beck, nature enters the political sphere wherever it is publicly defined as constituting a risk, yet risks are precisely those things that cross and dissolve the previously secure boundaries that defined the political sphere. Thus, it is not accidental that whenever environmental groups raise a concern about nature, it is almost inevitably defined in terms of (usually social) risks and in combination with an attempt to indicate political responsibility. Beck, thinking in systemic terms, regards the political mobilization of risk as a reaction to changing environmental circumstances that themselves have previous political causes, that is, as a kind of political feedback loop. The advantage of this view for political ecology is that green issues stop being matters of special or limited interest and take center stage as debate over the definition, extent, and provenance of risks proliferates. Such debates can perhaps extend as they gain public visibility into a critique of modernity as a whole. An optimistic reading might see this kind of mechanism, together with increased awareness and reflection on risks in general, as acting as a kind of political moderator on a global scale, like the governor on a steam engine, limiting excessive "acting into nature" and eventually generating a fully reflexive modernity. A pessimistic reading would regard it as merely waving while drowning in response to what has become a runaway process.

Beck hovers ambiguously between such optimism and pessimism but clearly wishes to retain something of environmentalism's critical potential to catalyze sociopolitical change. The problem is that reducing environmentalism to a processual side effect undermines the ethicopolitical concerns motivating many activists. The fact that, from Beck's perspective, there is no apolitical nature left to save does not just suggest that all environmental issues are politically contestable, which seems obviously true. It also seems to mean that the whole idea of nature as something other than ourselves about which we might be ethically concerned vanishes: "Modernization has *consumed and lost its other*" (Beck 1992, 10). Environmentalism becomes just another way of voicing human concerns about human health, prosperity, and so on, one that is conceptualized entirely in terms of risks to human society. In other words, that which begins as a radical critique of our ethical and political relations to nature ends with nature being entirely sidelined, its place taken by risk and reflexivity as the new "objects" of social concern, something exemplified by Beck's own later work, in which environmental concerns largely disappear. The loss of this ethicopolitical aspect is compounded by the treatment of responsibility, which Beck reduces to a now almost impossible attempt to locate a *cause* for environmental ills, again stripping it of its ethical motivations (Beck and Willms 2004, 118–19).

This is not to say that Beck dismisses ethics. He suggests that reflexive modernity ushers in a *"new ecological morality"* (Beck 1992, 77), which emerges as part of the political dynamic following the recognition of risks. But this morality only incidentally concerns actual harm done to nature or, for that matter, other people. What matters is not whether ecological or health effects are real but their risk perception: "If people experience risks as real, *they are* real as a consequence" (77), that is, they have political (and moral) reality effects. So understood, risks provide loci around which a collective politics can cohere, a politics that may well include normative values, but they do this, Beck argues, because they now constitute a perceived threat to individual or group interests. The ethics of reflexive modernity, then, is grounded in a recognition that we are all, each one of us, equally at risk (a dubious egalitarianism that has been the focus of many critiques of Beck's work). These ethical values emerge from the political recognition of shared self-interests in ameliorating perceived risks rather than actual concerns for others. Not only are ethics thus surreptitiously reduced to and redefined as a form of self-interest but they are also, simultaneously, sociologically reduced to their role and functions as just another mode of reflexivity within late modernity.[4] What is more, ethics seems to be regarded as a particularly ineffective kind of political reaction.

In Beck's words, it "is impossible to even imagine mobilizing an ethical movement that could oppose the global dynamic of unfolding technology. In this context ethics is like putting a bell on a 747. You can tinkle warnings all you want and technology will continue roaring ahead on autopilot" (Beck in Beck and Willms 2004, 204).

Leaving aside the fact that Beck's own preferred way of holding technology responsible, via scientists inexplicably coming to accept the programmatic uncertainty of their own research (205), hardly seems more promising; such statements are indicative of the way Beck misconstrues ethics and the ways it can inform political power (think, for example, of the fall of the Berlin Wall). His theoretical framework dissolves precisely those concerns about our ethical responsibilities toward nature and other people which comprise the core of environmental politics, interpreting them as merely one more mysterious form of reflexive reaction. But as Berking (1996) indicates, ethics and environmental politics are different in several ways from other forms of social readjustment. First, in being other directed, ethical responsibilities clearly bear little resemblance to narrowly defined interests or technological concerns about safety. Second, they both depend on and incite deeper, more radical, analyses of the responsibilities associated with acting into nature, analyses that call for "creative change" rather than merely "instrumental adjustments" (189). Third, "institutionalized forms of political and economic action lack what distinguishes in particular the world-view structures of the social actors and environmental activists: a normative framing; that is, a *moral consciousness*" (190).

These distinguishing features of ethical relations are lost in Beck's analysis, which also fails to provide convincing accounts of how ethical values emerge and why ethical responsibilities are constitutive of individuals or why they relate to specific aspects of the environment. These theoretical gaps are not just oversights but consequences of both shifting the "object" of analysis away from that nature which, according to his theory, can no longer exist as society's "other" and of working with a conception of political action that has a denuded understanding of its intimate relation with ethics, especially in terms of responsibility. To the extent that Beck successfully shifts attention away from questions about nature (or for that matter, other people) to risks per se, ethical relations are not so much explained as explained away. After all, it is difficult to envisage having an ethical relation to (a concern for rather than about the consequences of) a risk qua risk. It is, however, in precisely such concerns for others that responsibility, as something other than a simple attribution of causation, or liability, lies. Beck, then, seems to expand the role of political action, reconfigured as reflexivity, into nature only by dissolving what, for Arendt, constitutes the entire rationale behind her own conception of politics, namely, drawing together the intimate connections among individuals, their actions, and ethical responsibilities.

This shift in attention and its consequences are, as might be expected, paralleled in Latour's descriptions. Conceptualizing the world in terms of hybrid networks rather than the modernist separation between political subjects and apolitical natural objects reveals their interdependency. But it also means that the "political ecology" Latour endorses often has little or nothing to do with saving nature in terms that environmentalists might recognize. For Latour, the place of natural beings in environmental politics is taken by "matters of concern" that are politically and ethically constituted by both human and nonhuman actants. Moralists, described by Latour as the ethical equivalent of the scientific expert, have the job of shuttling back and forth across, and *unsettling,* the provisional boundaries set up by any political constitution, constantly reminding everyone of what has been left out or excluded from the equation. This has a certain plausibility insofar as it might be regarded as a posthumanist equivalent of the infinite questioning that constitutes ethics for Levinas and Murdoch (see chapter 2). It is also relatively easy to connect this analysis with Agamben's criticisms of the (a)political constitution installed by the anthropological machine—especially because Latour also explicitly uses the term *constitution* to refer to the metaphysical "division of beings into human and nonhuman, objects and subjects, and to the type of power and ability to speak, mandate, and will that they will receive" (Latour 2004, 239). However, this still leaves key questions unanswered. Most especially, *who* are these "moralists" who are sufficiently motivated to engage in such practices? *Who* actually cares?

To say these moralists are actants involved in matters of concern may help reconceptualize ethical relations in terms of their modification of and by other actants in hybrid networks. In this sense, it gives recognition to the active roles of more than just human individuals. For example, referring to Murdoch's kestrel (see chapter 2) as an actant *might* help envisage how the various players in that situation were involved in very different ways with each other in initiating an ethical event. It may, as Whatmore (2002, 166) argues, help us understand how ethical considerability can easily be dispersed "beyond the unified (and always) human subject" that typifies humanism and also serve to "complicate the bodily distribution of ethical subjectivity." But then, one does not need a notion of hybridity to accomplish either of these things; a little environmental ethics and an interest in phenomenology will do. More importantly, by itself, it says little about who is involved in expressions of ethical concern, either in terms of that particular hovering kestrel or Iris Murdoch. It deflects rather than makes an effort to answer Arendt's question by redescribing *what* the components of an ethical situation are and also what the role of a moralist is, or should be. But that which delimits ethics as such and every ethical relation is, as Levinas and Murdoch argue (see chapter 2), precisely its uniqueness, its singularity. A matter of (ethical) concern always involves a particular being who is capable of a moment of "unselfing," of suspending, or being suspended from, her self-referential obsessions, the world-distorting influences of self-regard, in coming to regard others as singular beings. Without beings like Iris Murdoch who are capable of experiencing such moments of suspension, there would be no ethical relations in the world at all. And, of course, without beings such as that kestrel, there would be no "others" to be regarded as if they were an end in themselves—and again there would be no ethics in the world. So if ethics is not to be dissolved into some postmodern hybrid equivalent of pantheism in the form of, say, cybernetic flows of information, it has to be understood in terms of its individuated phenomenology, its relation to an individual's feeling, thinking, and acting with concern for others, and in this sense the delimiting aspect of ethical experience (that which ensures that there is a possibility of ethics *as such* in the world) is *who* cares. All that an ethical posthumanism needs to claim is that this "who" is neither coextensive with nor solely concerned with those categorized as (properly) human, that there are no prepolitical Natural limits on who is concerned in either sense.

In other words, ethical action delimits and is delimited by *who* those expressing (or repressing) ethical concern are, their life histories, their decisions, the paths they took, and so on. It is precisely not an overdetermined response to a given environmental situation but an underdetermined responsibility emerging within a singular condition of undecidability that has been a lifetime in the making. To speak only of a matter of concern, as if it somehow composed itself *as a concern,* is to excise the who-cares whose ethical concern is the heartfelt singular condition of an event wherein their own self-interested possibilities are momentarily suspended. In other words, if politics as such is the condition of undecidability of pure means, ethics as such is the condition of undecidability of impure ends—it is the condition of responsibility, of being able to be concerned for others as if they were singular (unique and infinite) ends in themselves. And such moments of unselfing, whether consciously decided or not, are still, more than any other aspect of our existence and quite contrary to all self-centered (or moralistic) expectations, always the events that provide the most revealing instances of who we are. That is why her response to the kestrel expresses something about Iris Murdoch as a person.

A danger with Latour's focus on matters of concern (as with Beck's focus on risks) is that while the composition of these fuzzy new "entities" may indeed be deserving of theoretical attention, what begins as an antimetaphysical posthumanism is easily extended into an unjustifiable (unethical and apolitical) reduction of human beings to a post*human* condition—a condition in which the ethical and political openings that are largely (if not exclusively) initiated because of the presence of individual human beings are discounted to the detriment (even destruction) of ethics and politics as such. *The danger, once again, is of reducing human beings to bare life*—not in terms of an arbitrary sovereign decision, although there is a certain theoretical fiat involved here, but by a refusal to recognize the vital importance of certain dimensions of undecidability for certain beings, including the condition of possibility of each individual making ethical and political "decisions" for themselves. In other words, the danger is that of a biopolitical reduction by a redescription that treats people as if they were nothing more than their roles as actants and whereby other entities too become mere resources for the composition of matters of concern, of no interest in and of themselves.

This may not be Latour's intention, but it is a danger inherent in taking his claims too literally and/or of taking *his* account of ecological concerns as all that matters about them. But, while he offers a novel description of what goes on in ecological ethics and politics, an interesting posthumanist story about the constitution of "matters of concern," this story cannot replace or stand in for ecological ethics or politics as such. Describing something as a matter of concern cannot replace actually being concerned about something, since the plausibility of the former description is entirely parasitic on the existence of the latter situation. And this situation must include individual beings who are so concerned. In this sense, and despite explicitly claiming that his political ecology leads to a *"liberated"* rather than a liberal state (Latour 2004, 206), Latour's replacement of expressions of real concerns for nature by vicarious redescriptions of these concerns echo, in a tellingly ironic manner, both the language of scientific detachment and the liberal idea(l) of the ethically and politically neutral state, that is, the very institutions he claims to unsettle. It is otherwise hard to understand why readers would have to look so long and so hard to find anything resembling an expression of actual ethical or political concern for the more-than-human world (or even for other human beings) in Latour's *Politics of Nature,* still less any overt statement of support of the aims of actual political ecologists.[5]

To avoid very real biopolitical dangers, any posthumanism has to resist the temptations of humanist metaphysics (the anthropological machine) *and* resist reducing the dimensions of ethical and political action to a spurious equivalence under a generalized notion of modification—especially one subject to definition by "experimental protocol" (see earlier in this chapter). In other words, it is necessary to understand the differences between various kinds of actions that resemble each other only in terms of a very extended "family resemblance" (Wittgenstein 1988). From an Arendtian perspective, Beck's work in particular separates action from its ethicopolitical connotations by reducing it to an automatic reaction that does not require a moral consciousness, the thoughtful concern for others that Latour too passes over. Beck does this because he actually accepts the idea that society and nature are becoming a unified *process.* In other words, Beck not only *describes* the systemic processes that he claims constitute risk society but also builds them into his own theoretical account in such a way that they appear as inescapable. The kind of world alienation that Arendt is concerned to critique actually becomes the theoretical premise of Beck's work, a key part of the contemporary human condition wherein "everyone is cause *and* effect, and thus non-cause" (Beck 1992, 33). But for Arendt, the very essence of political action is tied to the human ability to initiate change, whereas "it is in the nature of the automatic processes to which man is subject, but within which and against which he can assert himself through action, that they can only spell ruin to human life" (Arendt 1993, 168). Environmental *activism* is, in this Arendtian light, an attempt to initiate political change against what are portrayed as automatic processes.

Of course, it might be unfair to describe Beck's concept of action (reflexivity) as automatic when "potentially chaotic" is more accurate. There is no prescribed direction that reflexivity has to take. Beck's systemic processes do not have the totalizing or universal coherence associated with the ideas of history or nature as process by their proponents (or by their critics like Arendt) with the ideas of history or nature as process. Unpredictability and instability ramify systemically throughout risk society, breaking down once relatively self-contained fields of activity, the family, class, nature, and so on, in a process of cascading and accelerating reflexivity. Everything links to everything else. For this same reason, Beck is generally critical of systems theories that posit external, interlocking, relatively self-contained and irresistible social processes over and against which the individual is virtually powerless. Systems theory has, he claims, been "thoroughly refuted" (Beck 1998, 37). But Beck still describes the dissolution of these fields and systems *systemically,* as occurring because of, and as part of, a *process* of increasing reflexivity, which itself constitutes both a continuation and a break with (that is, a new form of) modernity. This might explain some, though not all, of the contradictions apparent in his position, which, positing global systemic reflexivity, cannot therefore avoid thinking in terms of process and system to describe and explain risk society. Thus, in a particularly telling example, Beck (1992, 33) describes our environmental predicament as one which

> reveals in exemplary fashion the ethical significance of the system concept: *one can do something and continue doing it without having to take personal responsibility for it.* It is as if one were acting while being personally absent. One acts physically, without acting morally or politically. The generalized other—the system—acts within and through oneself: this is the slave morality of civilization. . . . This is the way the "hot potato" is passed in the face of the threatening ecological disaster.

It is genuinely difficult to tell here whether Beck is describing the current ethical situation or criticizing a particular interpretation of that situation that regards it as a systemic problem. The context, within the introduction to *Risk Society,* certainly suggests the former, and the moralizing tone, the latter. As already indicated, Beck often refers to how the complexities and interrelatedness of all aspects of society

and nature are said to make determining questions of ecological responsibility (qua origin and causation) impossibly difficult. But, within
Beck's theory too, responsibility is a hot potato distributed throughout
society alongside his all-pervasive risks, its "heat" only becoming apparent (part of the political space of appearances) through the collisions of expert discourses, lay opinions, and potentially catastrophic
environmental events. But this moment of appearance is precisely
where he has to reintroduce the political in a more Arendtian sense,
that is, as a space of individual and collective action, even going so far
as to argue that reflexive modernity represents a renaissance of political subjectivity (Beck 1997, 102).

Risk requires *political* solutions, and these solutions only come about
through the creative innovations of actors, of political individuals and
collectivities, for example in terms of new environmental social movements. Not only "social and collective agents, but *individuals* as well,
compete . . . with each other for the emerging power to shape politics"
(Beck 1997, 103). But this, of course, seems to entirely contradict a
situation in which no one acts politically or morally because it clearly
suggests that individuals *are* capable of initiating political change and
that, at least to some extent, they might also be willing to take responsibility for their actions in so doing.

Unfortunately, what Beck gives by slight of one theoretical hand,
he tends to grab back with the other, for these agents are themselves
produced through new processes of individualization by the systematic
denial of systemic responsibility! What this means is that risk society
is characterized by the paradoxical existence of an individual who,
having been cut adrift from the relatively secure grounds of existence
in industrial society, one in which family, class, and so on, provided
a common ground for identity and values, has to become the "*actor,
designer, juggler and stage director* of his own biography, identity, social networks, commitments and convictions" (95). Individuals do this
not by dint of making free, "existential" decisions but under the compulsion of contemporary social conditions. The individual is "kindly
called upon to constitute himself or herself *as an individual,* to plan,
understand, design and act—or suffer the consequences which will be
considered as self-inflicted in case of failure" (97).

Individuals become little more than "biographical solutions of systemic contradictions" (137) as the process of individualization throws
all responsibility back onto the subject despite (or rather because of)

the fact that risks are systemically produced. It might be said that where the systems theory description of "old" modernity allowed one to say that no one was responsible (it is the system's fault), the systemic unity of reflexivity allows Beck to say that everyone is (supposed to be) responsible (guilty), just as every individual is equally at risk. But, as already noted, (see chapter 3) Arendt argues that this will not do. Politics and ethics both require that we make judgments about and take individual responsibility for our actions.

For Beck, the individual in risk society is forced to take responsibility as a side effect of systemic failures so profound that any idea of "the system," as something that retains a coherent identity, that could itself be isolated and held responsible for its activities, becomes valueless. The side effect becomes "the motor of social history" (Beck 1997, 32). This is certainly contentious on a number of levels, not least because it seems to make reflexivity both omnipresent and omnipotent, the at-once so very nebulous and yet all-pervasive and inescapable feature constituting the human condition in late modernity. This ubiquitous presence perhaps explains why, for Beck, there is no alternative but for human beings to become increasingly reflexive, by which interestingly and despite previously blurring the difference between the two, he actually seems to mean more reflectively aware of our (reflexive) condition. Why this should help is by no means obvious unless one accepts the efficacy of thinking as a mode of ameliorating rather than exacerbating side effects. In other words, he actually seems to need to recognize that reflective thought is not the same as, nor can it be reduced to, just another form of reflexivity—otherwise the knowledge that we live in a form of reflexive modernization would itself be entirely useless. Whether he can recognize this difference without dissolving the theoretical coherence of the very concept of reflexive modernization is a moot point.

For Beck, responsibility is, one might almost say, *dumped* on individuals in the sense that subjects are forced to biographically recreate themselves in a vain attempt to relate their actions to totally unpredictable and contradictory systemic events. On the one hand this seems to mean that, from Beck's quasi-functionalist position, there are few actual grounds to distinguish the individual's adoption of these responsibilities from those grudgingly accepted by commercial companies, agencies, and so on. Such companies are forced by the proliferation of risks to show that they too are acting responsibly with regard to their products. This is the "source of the 'new piety' of business: ecological morality, ethics and responsibility are proclaimed for public relations effect" (Beck 1997, 128). Ironically, Beck argues, companies evade ultimate responsibility by portraying themselves as acting in the most responsible ways. Beck's skepticism about business motivations are no doubt well founded, but once again there seems to be an ethical cynicism at play in Beck's work that also reduces individual and collective responses to the equivalent of a form of self-interested structural adjustment. Only the fact that individuals can apparently still *create* their own biographies seems to leave some room for individuals' unique initiatives, for *acting* in the Arendtian sense, even if this is more a matter of saving one's public face rather than actually taking public responsibility for one's actions and thereby revealing *who* one is.

Arendt offers a more radical understanding of the relation between risk and responsibility because she retains the individual's potential for initiating political actions, for acting into the political sphere as such rather than just composing his or her political autobiography. The *responses* made by individuals to political circumstances, which today include ecological circumstances, are not the same as the reflexive reactions of the "natural" and/or nonindividual agencies, even though these too are involved and implicated. This is not (just) because of a difference between being able to reflect on one's actions and reacting automatically. Commercial agencies, for example, are clearly able to calculate costs and benefits and behave accordingly, and individuals often respond without thinking things through: they act spontaneously or in accordance with their previously established character. The difference lies in the existence, although not necessarily the deployment, of the individual's ethical conscience—of the individual being someone *who cares,* and not just something that reacts.[6]

This awareness is what links the ability to respond and ethical responsibility in every individual's political acts. From Arendt's perspective, to be a person is to be able to initiate actions and also to recognize and be concerned about the fact that all such actions will necessarily impinge on others. Acting is precisely not playing a role in which, to use Beck's words, it is as if we were "personally absent." Acting is the articulation of people's presence, of being persons whose multifarious identities are expressed and confirmed through showing themselves in word and deed in concert (and sometimes in conflict) with others. It is in the plurality of politics, this expression of difference from, but not of indifference to, others that responsibility as more than mere causation lies. Certainly it is associated with the political initium, the creative origination of events by persons, with the natality of action, the second birth of the individual through political enactment (Disch 1994, 32). But, because it is personal, political, and ethical, responsibility is never merely a matter of locating or attributing a prior cause.

From Arendt's position, too, there is nothing new about the unpredictability of one's actions, the fact that they will have unforeseen and incalculable consequences. This is a defining feature of action that is always inherently unstable in the ways it plays out and always prey to differing circumstances. To be a person is to accept this and still be able to respond, to act, in such a way that one recognizes and remains concerned about the effects that one's actions may have on others. This remains so even in the most constraining of political circumstances, even, as Arendt makes clear, under a dictatorship. To suggest that we are now, all of a sudden, incapable of initiating actions, bearing responsibilities, or, in short, of being persons because we are beset with potentially disastrous environmental problems is itself an abdication of responsibility on the grand scale. Environmental activists disprove this thesis daily, and they do so because they do not regard their actions as inconsequential bell-ringing or themselves as just a concatenation of side effects. They do so because they accept some responsibility for our situation (they take it personally) and believe, quite rightly, that their actions, their creative responses, can and do make a difference.

For all its complex nuances and the importance of his insistence on the intertwining of nature and politics, Beck's account fails to take the nature of political action and environmental responsibility seriously, reducing both to process. To be sure, this process is no longer envisaged as linear or predictable, but it nonetheless retains a hidden power over the appearance of nature and human action. In other words, rather than expanding politics, the space of appearances, into nature, Beck makes both politics and nature subject to hidden systemic processes. In terms of Arendt's (1958, 296–97) description of modernity, in "place of the concept of Being we now find the concept of Process. And whereas it is in the nature of Being to appear and thus disclose itself, it is in the nature of process to remain invisible." Hidden powers, whatever form they take in different societies—fate, history, or in risk society, side effect—can always deflect attention away from the possibilities opened in individual responsibilities and actions.

Ecological politics long ago recognized that nature is part of the contemporary space of appearances. Through politicizing our relations to nature (but not installing another Nature), we ensure that environmentally irresponsible actions come to light and that the responsibility for the current distribution of risks, which is far from egalitarian, does not fall equally on all. Most importantly, ecological politics offers an alternative to a continual spiral of reflexivity by exposing its inherently political nature, that is, the fact that it is not an inescapable process. For nature to become political in an Arendtian sense cannot mean that its reality and appearances are subsumed within some hidden systemic process any more than human politics entails that other human beings become subsumed within and by a hidden political process. Politics drags all hidden processes kicking into the open, exposing their claims to be the inescapable bedrock of the human condition. In becoming political, nature is not eroded, but it too is recognized in its plurality and its natality. Acting into nature must, like acting into the political sphere, involve responsibility for others, concerns about effects, and making choices. It is precisely where such choices do not exist, where activities are initiated under the guise of being inevitable or automatic, that their political nature needs to be revealed.

We act into infinite possibilities, and yet we are responsible for our acts. Although played out in many different forms, this is always the paradox ethics presents. And in this we are aided only by what wisdom we might have gleaned from the tree of knowledge and encouraged only by that which emerged last of all from Pandora's box, by the hope that speaks of other possibilities. As Arendt (1993, 170) remarks:

> It would be sheer superstition to hope for miracles, for the "infinitely improbable," in the context of automatic historical or political processes, [but] . . . historical processes are created and constantly interrupted by human initiative, by the *initium* man [sic] so far as he is an acting being. Hence it is not in the least superstitious, it is even a council of realism, to look for the unforeseen and unpredictable, to be prepared to expect "miracles" in the political realm. And the more heavily the scales are weighted in favor of disaster, the more miraculous will the deed done in freedom appear; for it is disaster, not salvation, which always happens automatically and therefore must always appear to be irresistible.

6 ARTICULATING ECOLOGICAL ETHICS AND POLITICS

If ethics without politics is empty, then politics without ethics is blind.
—Simon Critchley, *Infinitely Demanding:*
Ethics of Commitment, Politics of Resistance

HOW MIGHT THE RELATIONS between political action and eco-logical (ethical) responsibility begin to be envisaged in such a way that each informs the other and yet neither is made subject to the other? How do we dissolve the claims of sovereignty and yet retain a poli-tics informed by the Good where each is understood as an expression of natality, diversity (plurality), and as exemplifying the appearance of those individuals who feel, speak, and act? Another way of posing the question would be to ask whether an ecological ethics might come to delimit, but not dictate, how political communities choose to act into the world.

There is an understandable but regrettable tendency in environ-mental ethics to translate ethical concerns for nonhuman others into more or less fixed moral frameworks—often on the basis of naturalistic claims about supposedly objective (intrinsic) moral value attaching to certain species (Smith 2001a)—values that are then regarded as impos-ing constraints on the freedoms of human-centered politics. To take just one example, and there are many, Laura Westra (1998) develops a notion of ecosystemic "integrity" as a supposedly measurable "objec-tive state" (9) equivalent to the "optimal functional capacity" (241) of that system, an optimum (always) achieved through its evolutionary (natural) development. Integrity should also, she claims, be understood as a primary moral good threatened by human intrusions because "anthropogenic stress" leads to suboptimal ecosystem functioning and "non-evolutionary changes" (100) that constitute a moral wrong to that ecosystem. This proposal is questionable on many grounds, including ecological science (see Smith 1999b), but the key problem is that, having made what she believes to be a convincing philosophical argument concerning the Good of ecosystems, and having translated her concerns into a universal principle *(archē)* of integrity, Westra's (1994) conclusions are then supposed to impose specific and extensive limits on political action. "The first moral principle is that nothing can be moral that . . . cannot be seen to fit within the natural laws of our environment in order to support the primacy of integrity. . . . Act so that your action will fit (first and minimally) within universal natural laws" (Westra 1998, 11).

What we are presented with is thus another version of the sovereignty of nature, which simultaneously reiterates Plato's arguments concerning the sovereignty of the (in this case ecosystemic) Good over the polity. Westra (1998, 150) is quite explicit about this, arguing that "in contrast to the tenets of 'political correctness' and individualistic modern liberalism, we may be able to cast some serious doubts on the capacity of our institutions, at least in their present form, to do better in both theory and practice than the Platonic philosopher." And while Westra recognizes that there can and will be debate among ecological scientists, philosophers, and other stakeholders concerning the definition and analysis of integrity, the "ultimate reality of the concept (as definable, quantifiable, and applicable), and hence its validity in both law and morally, need not be questioned" (10). But one does not need to be either a liberal or an advocate of contemporary institutional structures to see the antiethical, antipolitical, and even potentially totalitarian consequences of a move to impose "a common conception of the 'good,' that is not open to revision and rejection" (150).

Even if Westra's ire is targeted primarily at making ecological decisions according to utilitarian and majoritarian calculi of "democratically supported preferences" (10), it is ethics and politics *as such* that her system subjects to the supposedly politically impartial concept of ecosystem integrity.[1] This claim to be above politics excludes this sovereign principle from ethical and political questioning even as it places those deciding when and where it should apply in a position of absolute and universal authority over ethics and politics as such. That Westra's ecological and philosophical stakeholders are more nebulously defined than Plato's philosopher-kings (although Westra indubitably places herself in this position of privileged interpreter [steward] of nature's sovereign principles) makes this biopolitical system more, not less, mysterious and open to abuse. That Westra (1994, 200), rather than supporting individual state sovereignty, advocates a "'world order' institution" capable of enforcing the policies *demanded* by the principle of integrity hardly ameliorates such concerns.

Neither environmental ethics nor ecological politics should be understood in this way, and not just because the approach Westra typifies is indicative of a moralistic antidemocratic authoritarianism (Dobson 1995, 80–85) or because it invokes the naturalistic fallacy— taking what *is* (purportedly) the case for what *ought* to be the case—or even because it fails to recognize the social and historical particularity of ideas of nature (Soper 1995; Smith 1999a; Castree and Braun 2001), including the very idea of an ecosystem (Golley 1993), but because it rests on a restrictive misunderstanding of ethics and politics as such. Lacking this understanding, it fails to recognize how an ecological ethics might affect/effect politics without simply making politics subject to it, without, for example, moving straight from principles, such as those of ecosystem integrity, to governmental policy and biopolitical forms of managerialism.

As already argued, ethics and politics as such, and the relation between them, should be understood anarchically, that is, in terms of the rejection of any principle *(archē)* of sovereignty altogether. And one way of attending to the myriad possible paths among ethics and politics as such might be by elucidating further connections between Levinas's understanding of ethical responsibility (and ethics as such) and Arendt's understanding of political responsibility (and politics as such; see chapters 2 and 5). The emphasis in what follows is on Levinas rather than Murdoch, first because despite the (metaphysical and ecological) drawbacks of Levinas's philosophical approach (see chapter 2), his work provides a much more detailed account than Murdoch's ever does of just how ethical responsibility arises; and second, because there is already a body of secondary material that has attempted, however unsuccessfully, to draw out the political implications of his ethics. Unfortunately, although successfully undercutting the claims of sovereignty within ethics, especially in relation to the (supposedly sovereign) individual, Levinas too tends to fall back into this same language of (individual and state) sovereignty where politics is concerned, and this is because he lacks the kind of understanding of politics as such that Arendt provides. Understanding this failure, its ecological implications, and its possible solution requires reiterating and expanding on some of the core aspects of Levinas's and Arendt's positions.

Politics and Ethics As Such

> That man is more of a political animal than bees or any other gregarious animals is evident.
>
> —Aristotle, *The Politics*

Most people influenced by the Western philosophical tradition would concur with Aristotle's statement, for humans are, as he argued, *bios politikos,* beings who constitute themselves as (differently) human within and through their political forms of life. But this is not, as Hannah Arendt (1958, 12–13) points out, an attempt to define humanity as a particular kind of biological species (an early iteration of the anthropological machine) so much as a claim about the importance of sustaining a political life for the expression of human freedoms.[2] Politics is, in this sense, the worldly medium of human beings' different possibilities, of how, in certain circumstances, they can come to freely express themselves as themselves, as individuals and not simply as beings constrained by biological necessities (by their work) or as functionaries fulfilling productive roles dictated by social conventions (by their labor). Remember, for Arendt, the words and deeds that comprise (political) action express *who,* not *what,* we are (see chapter 5). As Giorgio Agamben (2000, 3) argues, the *bios* in *bios politikos* refers to a "form of human living [that] is never prescribed by a specific biological vocation, nor is it assigned by whatever necessity; instead, no matter how customary, repeated, and socially compulsory, it always retains the character of a possibility; that is, it always puts at stake living itself."

This is why politics, on this reading, is not at all a *means to an end,* especially not an end predetermined by some essential human nature, whether selfless altruism, selfish genes, or selfish instincts, as so many political theorists have argued. Politics is, in Agamben's (2000) terminology, a "means without end," a pure means, a practice valuable not for what it produces but only insofar as we value human freedom itself. We (that is, those who are so concerned) must also recognize that there are consequently no guarantees that human lives will always have a political aspect, since this is not a fact of nature but something that we, through our own actions, must strive to sustain. Such freedoms are all too easily lost. As Agamben (1998) argues, the principle used to justify exercising political authority over others, the principle of sovereignty, is precisely a claim to be able to strip those subject to it of their political possibilities, their freedom of expression and association, to reduce them to the condition of "bare life."

This understanding of politics as a tenuous but vital freedom of self-expression through words and deeds in the face of others, as an opening on the infinite possibilities of being (differently) human, seems far removed from those practices often labeled "politics." It might also seem unduly utopian in many respects, not least in the sense that action is never entirely freed from the requirements of work and labor or from our other necessary involvements in nature and culture. But this is, one might argue, how we might delimit politics as such, as an event that, while dependent on, is nonetheless irreducible to, work and labor. And the value of this understanding can be seen precisely because every other conception of politics as a (relatively) autonomous practice, and every attempt to claim political authority, is parasitic on this prior understanding—an understanding, we might say, that comes before, or better, lies *beyond,* all of its constitutional manifestations. The constitutive power (the associative potential) of politics as such to create a community in and through the differences between us lies beyond, even as it underlies, attempts by political authorities to constitutionally control, define, channel, utilize, and constrain the power of "free association."

All this has already been said, but clearly one implication of this understanding is that the state (polis) is not, as Aristotle (1988, 3 [*Politics,* 1253a ln.2]) claims, "a creation of nature"; rather, it is revealed *before all* as a *political* creation, even though every state retroactively tries to naturalize itself, to reimagine itself as the natural repository of all (authorized) political action. Aristotle (4 [*Politics,* 1253a ln. 25–27]), of course, argues that the "proof that the state is a creation of nature and prior to the individual is that the individual, when isolated, is not self-sufficing: And therefore he is like a part in relation to the whole." But if this insufficiency is understood just as a matter of *survival,* then bees would be no less, and given their degree of dependence and "social" integration, perhaps *more* political than humans, which is quite contrary to Aristotle's intent. Once again we see that the political *life,* the *bios politikos,* cannot be simply a matter of an individual's capacity to survive or not: it is not reducible to bare life; rather, it has to do with having the potential to initiate, participate in, and sustain politics as such. And this requires, Aristotle argues, a "sense of good and evil, of the just and unjust and the like, and the association of living beings that have this sense makes a family and a state" (4 [1253a, ln. 16–17]).

This understanding of politics as such has much in common with, and in many senses complements, Levinas's understanding of ethics as "first philosophy," as a regard for and an infinite responsibility toward Others that emerges through our face-to-face encounter with these (differently) human individuals. I argue that, in many respects, Levinas's understanding of ethics is not only compatible with but could be considered *constitutive* of such an understanding of politics, that what Arendt refers to as the "in between'" of politics as such is, at least to some extent, dependent on the "between us" (Levinas 1998) of ethics as such, and vice versa. This, of course, is a much stronger claim and has to be interpreted carefully to avoid any temptation to regard ethics or politics as such as either equivalent, reducible, or in a hierarchical relation to each other: they are not. However, Levinas's understanding of ethics as such is complementary to an Arendtian take on the political in several ways, all of which relate to their attempts to articulate this "as suchness," the singularity of these relations that always lies before, beyond, and still exceeds its subsequent crystallization in, for example, a body of moral or legal codes or a particular political ideology. They each understand ethics or politics as such as *constitutive* of human associations and of who we are as (differently) human individuals, and only secondarily, if at all, as *constitutionally* defined limits on such expressions: they emphasize the anarchic aspects of ethics and politics.

Levinas and Arendt: Anarchē, Ethics, and Politics

> The notion of anarchy we are introducing here has a meaning prior to the political (or anti-political) meaning currently attributed to it.
>
> —Emmanuel Levinas, *Entre Nous: Thinking of the Other*

Although Levinas often refers to his understanding of ethics as anarchical (for example, Levinas 1998, 99–102; 2003, 45–57; see also Greisch 1991, 80; Benso 2000; Abensour 2002), and the anarchical strand of

Arendt's work too has sometimes been noted (Isaac 1992), neither are anarchists in the usual political sense. However, we might say that ethics as such are anarchic in at least three inseparable ways. First, Levinas rejects the idea that ethics as such can be captured within or ruled over by any overarching moral principles and concepts, focusing instead on the interruptive power of the ethical event. This, I argue, links to Arendt's critique of the (political) dangers of simply adhering to dominant moral norms and also to that systematization (and bureaucratization) of the political sphere that seeks to replace political action as such with rules and processes. Second, and closely linked to this, Levinas claims that ethical responsibility arises without and before any definable point of origin; it has no *archē* (beginning) in the ontology of the world; it is beyond or otherwise than Being. This, I suggest, resembles Arendt's understanding of the "miraculous" initiating power she associates with politics as such, its natality—the liberating power to engender new beginnings despite what are often portrayed as the unchangeable "givens" of a social situation. Third, Levinas regards the ethical relation to the Other as Arendt regards politics, as a matter that "concerns the individual in his singularity" (Arendt quoted in Young-Breuhl 2006, 201; see chapter 2). The connections here are complicated, but the point is that political action and ethical responsibility are what constitute the individual as an individual and that ethics and politics *as such* are constituted through attentive responses to those Other individuals with whom we become associated.

The theoretical inseparability of these anarchic aspects makes it impossible to treat them in isolation. The issue of singularity inevitably introduces the question of ontology and also of the resistance of ethics and politics to inscription in rules and codes. Nonetheless, singularity offers an initial way to link Levinas's thought to Arendt's, since each argues that ethics and politics are initiated and sustained through intimate face-to-face encounters between singular individuals, encounters in and through which we come to glimpse something of *who* the Other/other facing us is without ever fully knowing her or him. Both also argue that such encounters require the acceptance of the Other's/other's radical differences from ourselves, a refusal to see them only (or at all, in Levinas's case) in the light of our desires (as fulfilling what *we* need or in terms of what they can do for *us*) or to subsume them under preconceived categories that reduce them to abstractions.

If Levinas emphasizes the transcendence of ethical difference, the

ways in which the Other precedes and goes beyond any egotistical desires, Arendt emphasizes otherness in terms of the necessary plurality
of politics. In each case, if everyone were essentially the *Same,* if we
were not singular individuals, there would simply be no possibility of
any ethics or politics as such. Nor, of course, would anything new, any
novel understanding, emerge from our encounters. The ethical and the
political spaces, where the Others/others express something of themselves as themselves, in their singularity, emerge only through creative
associations that concern themselves with sustaining such different/
plural possibilities.

These ethical and political associations differ insofar as Levinas is
concerned with the emergence of ethics as such, as a prior association
between self and Other composed as a "fundamental structure of subjectivity" (1985, 95). The encounter with the face of the Other evokes
a proximity that touches and troubles us before we can conceptualize
its effects on us, a trace that does not allow "itself to be invested by the
archē of consciousness" (Levinas 1996, 81) and which dispossesses us,
that is, draws us out of our self-possessive concerns. This, as we have
seen, is also Murdoch's claim. In other words, ethics challenges the
very idea of a sovereign individual with absolute authority to decide on
what does and does not concern him or her. The fact that this relation
is *pre*conscious also means that it necessarily transcends (lies beyond)
its inscription into the linguistic categories within which consciousness is formulated and that denote the accepted (ontological) order of
the world. "Anarchically, proximity is a relationship with a singularity,
without the mediation of any principle or ideality" (81).

Arendt, on the other hand, is concerned with the emergence of politics as such, an association between self and others in what appears as
a broader public realm, one where our relations to others are mediated
through *words* as well as deeds.[3] For her, words are an expression of
others' individual characters, but this does not mean that their public
persona encompasses everything there is to that person—just that this
is how they choose to appear politically, and it is this political appearance that concerns her above all else. For her, *appearances matter* quite
literally: that is, they have material (political) effects on others. But she
does not fall into what Levinas would regard as the trap of treating
the human individual as the pregiven (ontological) basis of political
intersubjectivity. The political self, the individual who does more than
just labor or work, is far from being a pregiven entity; indeed, individu-

als constitute themselves as themselves only within the possibilities offered by political space.[4]

The "who" of Arendtian politics is not equivalent or reducible to either the "I" or the Other of Levinasian ethics, although she or he may (indeed must), if Levinas is right, have been first constituted through exposure to such ethical relations. That is, individuals are never *simply* self-interested, nor are their actions, insofar as they are individuals, *simply* motivated by their need to compete in a struggle for existence (although neither Levinas nor Arendt would in any way deny that people often act competitively and selfishly—see later discussion). The ethical self and the political self are composed within and through what might be termed an *intimate ecology of responsibility* to others/ Others, a patterning of relations that the ethical/political self cannot avoid if she or he is to be some*one*.

Such responsibilities arise differently: for Levinas, one is held ethically responsible by and for the Other, by the face that *singles one out,* that addresses and contests one's identity. The experience of the face is one of being addressed by the Other, of passivity, and yet the "I" finds herself or himself bound by a responsibility for the Other that is absolute and infinitely demanding. This is an obligation that extends far beyond being responsible for my own actions toward the Other, even including a "responsibility for what is not my deed, or for what does not even matter to me" (Levinas 1985, 95). For Arendt, responsibility (which is ethical as well as political; see Assy 2008) arises as a direct result of her argument that political action reveals who the individual is. Insofar as the act I perform is incontrovertibly mine, since it is this action that marks *my* political appearance before others, then I alone can be held responsible for its consequences—this responsibility cannot be passed to anyone else. Yet the consequences of acting into the political realm are inherently unpredictable and continue to cascade forward into the future. In other words, the events set in motion by my actions are potentially limitless, and so my responsibilities too extend far beyond those associated with what I consider to be their immediate or intended consequences. These responsibilities too are, in effect, infinite and can be redeemed only by the possibility of others, who also understand the vagaries of the human condition, offering their understanding and forgiveness.

Importantly, Levinas's emphasis on the passivity, presubjectivity, and inescapability of the ethical encounter with the Other leads him to postulate ethical responsibility as an obsession, a compulsion, an absolute and infinite obligation. This is because ethical responsibility concerns a "subjectivity prior to the Ego, prior to its freedom and non-freedom" (2003, 51). For Levinas, freedom is associated with consciousness and hence the self-possession of the individual (1996, 82); it places a limit on responsibility in a way that is the antithesis of Arendt's perspective. Her emphasis on activity, emergent subjectivity, and the need to involve oneself politically—and to make judgments concerning, for example, the extent of that involvement (Arendt 1982)—sees the initiation of such responsibility in more voluntaristic terms. Freedom is absolutely key for Arendt; it is "the raison d'être of [Arendtian] politics" (Kateb 2000, 148). It is not easy to see how these two very different understandings are reconcilable, although at the very least, it provides another reason for recognizing the irreducibility of Levinasian ethics to Arendtian politics and highlights two different, yet potentially complementary, ways in which responsibility (for others) is ethically and politically articulated.

This also brings us to the issue of ontology: for both Levinas and Arendt, the singularity of the Other/other troubles the "given" ontology of the world. This is especially so because Levinas (rightly or wrongly) regards ontology as that dominant philosophical tradition which has concerned itself with making the world present to, and encompassed by, thought. Ontology considers all that there is, for example, of another person, as being inscribed within the phenomenal world of appearances and hence as potentially appropriable for our self-possessive purposes. This is another reason why Levinas describes the Other as beyond *being* (ontology), as always exceeding that made present in appearances, knowledge, or language. The Other who faces us ethically retains a capacity to surprise us, to interrupt the workaday self-centered world that would otherwise proceed almost automatically, without our intrusion and theirs.

Leaving aside potential philosophical differences between Levinas's and Arendt's understanding of ontology, both agree on the singular importance of not taking appearances for granted. Ethics and politics as such constantly introduce possibilities that escape or transcend what were previously taken as the ordering first principles *(archē)* of the natural or social worlds. And for this reason, among others, Levinas and Arendt are extremely critical of attempts to naturalize ethics or politics, to treat ethics as something that can be read off from, for

[handwritten margin note:] rejects sovereign move, calls his ideas anarchic, but not an anarchist

example, our biology (Levinas) or to treat human society as something that operates only according to pregiven structures and processes (Arendt). Ethical and political responsibilities cannot be derived from any fixed ontology of human behavior. Ethics as such is not reducible, for example, to a sociobiological understanding of altruism, to selfish genes. Indeed, Levinas explicitly "formulates his [ethical] thought as a radical alternative to social Darwinism" (Bernasconi 2005, 171). Arendt too targets both naturalistic explanations of and justifications for the sociopolitical order and is especially critical of the systematization of political structures that try to predict and eradicate political uncertainties, to get society to run like a well-oiled machine at the cost of politics as such. And this, after all, is precisely what biopolitics attempts to do—to treat politics as no more than a process to be efficiently managed.

What is important here is that both Levinas and Arendt recognize the vital importance of the ethical and political *event* that interrupts what would otherwise be taken for granted, what is taken as being the ontological order of the ethical and political world. For Levinas (2003, 32), "the face *enters* our world from an absolutely foreign sphere . . . exterior to all order, to all world." It is, in Murdoch's terms, "alien"—a visitation that disorients and challenges. Arendt too regards politics as such as an expression of that initiating and constitutive power that interrupts the given order of things, referring to this capacity to set events into motion as the "natality" of action, its almost miraculous ability to make new beginnings, to intrude on and change forever what had, up until that moment, been taken as natural, social, or historical necessities. "Beginning, before it becomes a historical event, is the supreme capacity of man; politically it is identical with man's freedom" (1975, 479).

This leads to the third anarchic aspect of Levinas's and Arendt's understandings, since the ethical and political event is a beginning, the principle and point of origin *(archē)* of which cannot be fully identified, encapsulated, or enunciated in language. In Levinas's terms, it is a *saying* (addressed to one or more others) that always exceeds and troubles that which is *said*—that which language serves to fix as the ontology, the totalizing *logos,* of the spoken or written wor(l)d. Saying is the anarchic expressivity that informs the spoken word, what is said, but is not thereby defined by that word (just as the singular Other comes from beyond and informs our being but resists being appropriated by it). "Saying resists becoming a theme . . . thematization makes every being into a said, i.e., into a being that is identified as a phenomenon within the context of a story or a discourse" (Perperzak 1997, 61).

This links to Arendtian politics in at least two important ways. First, Arendt emphasizes how singular expressions can interrupt the supposedly fixed flow of history. These expressions might be seen as exemplifying a political *saying* that informs, but also destabilizes, the overarching narrative themes of an otherwise depoliticized history (Smith 2005a). Second, she, like Levinas, emphasizes the importance of not reducing the (ethical or political) "as such" to just following rules or formulae. To accept this reduction is to replace politics with a kind of antipolitical and uncritical conformity with current norms, whatever they might be. This is typical of bureaucratic (and biopolitical) state regimes and in its most extreme forms is indicative of totalitarianism (Arendt 1975). This inattentiveness to individual ethical and political responsibilities is precisely what Arendt believes facilitated the easy transition from bourgeois respectability to acquiescence with the inverted (im)morality that marked the Nazis' accession to power. For Arendt, according to D'Entrèves (2000, 250), the exercise of political judgment requires "the ability to deal with particulars in their particularity, that is, without subsuming them under a pregiven universal." And this singular ability is also ethical in a Levinasian sense.

Levinasian and Arendtian Politics

These comparisons seem to bring us a little closer to elucidating how an ethically informed ecological politics might be conceived. It should at least be clearer how following the paths between Leviansian ethics and Arendtian politics avoids the quasi-Platonic pitfalls that reduce ethics to moral first principles *(archē)* and then treat these as sovereign over an ontologically defined politics—a fixed moral/political ordering of the world. Strangely, though, these paths between Levinas and Arendt seem little followed, even where human politics is concerned.

The relative dearth of secondary material directly comparing Arendt and Levinas's work is unfortunate and baffling, especially considering the shared historical context of their writings.[5] After all, their mature philosophies were both developed as a direct response to the rise of Nazism and its systematic reduction of Jews, Gypsies, and others first to bare life and then to dead matter in the "fabrication of corpses"

(Adorno in Agamben 1999, 81) in the camps. (Both Arendt and Levinas were also responding directly to Heidegger's active political support for Nazism.) Their focus on the anarchic aspects of ethics and politics as such became necessary because of the absolute and catastrophic failure of what had passed for ethics and politics in the totalitarian state of Germany and beyond (Bernstein 2002, 254).

Some might argue that this lack of comparative material also relates to real difficulties in relating Levinas's ethics to politics, especially the ambiguities that arise when trying to associate Levinas's philosophy with any particular governmental form or political ideology.[6] But such difficulties should actually make an attempt to connect his ethics with politics as such a more obvious path—which makes it all the more surprising that writers like Bergo (2003), who focuses explicitly and in detail on the political articulation of Levinas's ethics, hardly mention Arendt at all. Critchley (2002b, 1), who goes so far as to claim that "Levinasian ethics is not ethics for its own sake, but for the sake of politics, that is, for the sake of a transformed understanding of the organization of social life," belatedly suggests, at the conclusion of his *Ethics of Deconstruction* (1999, 237–38) that Arendt and Levinas might be fruitfully connected given their mutual interest in justice, and justice is certainly a theme taken up by others interested in connecting Levinas's ethics with politics (see Horowitz and Horowitz 2006) and by Levinas (1998, 195–96) himself. But even the tentative connections outlined here (their anarchic concern with singularity, ontology, the resistance to authoritative inscription, and the focus on complementary but irreducible aspects of individual responsibility) suggest that their philosophical, ethical, and political resemblances run much deeper than just a mutual concern with justice.[7]

Levinas's own attempts to find a place for politics alongside ethics falter largely because he never really develops a clear idea of politics as such, associating politics, especially in his earlier writings, with an almost Hobbesian war of each against all and a social-Darwinian "struggle for existence," which he regards as unethical not only because of the way this was directly employed by Nazi ideologues but because it explicitly reduces human life to a matter of mere survival, to Agamben's bare life. To oversimplify, the trouble is that Levinas accepts de facto that politics has always been, and can only be, biopolitics. For Levinas, the survival instinct is part of the ontology of our being and "the origin of all violence" (Bernasconi 2005, 177), and "war is nothing but the pure face of politics" (Dussel 2006, 79). Thus, at least in this (almost biologically) reductive sense, Levinas's understanding of politics is actually the very antithesis of Arendt's, and so the difficulty in connecting their thought is hardly surprising.

Arendt too recognizes that violence is a form of political action, an interruptive event, but specifically defines it as that which is employed to negate (constitutive) political power and is negated by it (Arendt 1970). War and violence constitute the limit of politics and the "justification of violence as such . . . [is] no longer political but anti-political" (Arendt 1965, 19). Both Arendt and Levinas often associate violence with attempts to enforce political authority over others, with establishing and policing a totalizing, sometimes an overarching totalitarian, social order that seeks to deny expressions of plurality and difference. But the question of using violence to resist violence is also, as Bernasconi (2005, 178) points out, a political problem and one that cannot be avoided. And so, even given his jaundiced view of politics, Levinas sometimes finds himself arguing that there can be an "ethical necessity" (1989, 292) as well as a practical and even military necessity for the defense of the state.[8] This seems difficult to reconcile with his view that "unfortunately for ethics, politics has its own justification," a justification that, taken to its extreme, creates a "direct contradiction between ethics and politics" (292), where "politics' own justification," for Levinas, is that of the struggle for survival, and ethics is an infinite, irrevocable, responsibility to the Other.[9]

This is why Arendt's much less restrictive and much more positive conception of politics, as a creative dimension exceeding bare life, where freedom and difference are manifested among "ordinary humanity" (Kateb 2000, 148), is so important. Certainly politics has, in Levinas's phrase, its own justification, but for Arendt, this is only in the sense that it is a means without end: its justification is certainly not success in the struggle for survival, nor are the freedoms it makes possible set in opposition to ethics.

As already intimated, and especially in his later work, Levinas develops a more complex account of the relation between ethics and politics in terms of justice. But Levinas's account of political justice is far from satisfactory because, as even he seems to accept, it works only to the extent that it does violence to, or at least compromises, his account of ethics. His account introduces an incorporeal "third" party behind self and Other, a "neighbor" who supposedly represents all others but can only do so in an abstract manner, in a relation that is not face to face and so, unlike the self's relation to the singular Other, requires linguistic mediation. The "third party looks at me in the eyes of the Other—language is justice" (Levinas 1991, 213). If this third party was just a figure (of the ethical imagination or of speech) intended to remind the self that ethical responsibilities are not contained within a relation to one singular Other, then this would not be so problematic. But Levinas clearly intends it as much more telling than this. Indeed, he argues, "The epiphany of the face qua face opens humanity" (213), which suggests that the third party *is* or represents humanity ("the whole of humanity that looks at us" [213]), or at least the other humans within the individual's ethical/political community. The emphasis on universalization suggests the first, that on justice within a state the second: Levinas equivocates between or conflates the two as and when it suits his inclinations.

But how can this be so? How can the third party be more than just a figuration of difference unless the self's responsibilities to all Other humans (or all the other humans in the self's state) are essentially the *Same,* something, after all, that Levinas's entire ethical oeuvre denies and something that simultaneously resuscitates the anthropological machine. How can the third party stand in for all parties without replacing the asymmetrical relation between the self and the singular Other with a symmetrical relation to *all* others and without turning an anarchical ethical relation into a relation contained within a political (and conceptual/linguistic) totality, a totality that, not incidentally, rules out all ethical and political relations to anything other than the human?

Perhaps Levinas's desire to speak to a secular politics framed entirely by and within the notions of humanism and state sovereignty led him to take this route. But whatever his motives, he certainly proceeds to link the notion of justice associated with the ghostly absence/presence of the third party in every ethical relation with a universalizing form of state politics: "In the measure that the face of the Other relates us with a third party, the metaphysical relation of the I to the Other moves into the form of the We, aspires to a State, institutions and laws which form the source of universality" (Levinas 1991, 300). And yet he knows that the consequence of treating all Others "according to universal rules, and thus in absentia," not in their singularity (300), is that it necessarily deforms the ethical relation. For this reason, too,

Levinas claims both that "politics left to itself bears a tyranny within itself" (300) and that "there is a certain measure of violence necessary in terms of justice" (1998, 105).

Even if we accept (as Levinas does and political anarchists certainly do not) the necessity of a state (qua a nation-state) rather than a polis in the broader sense of an associative political community, this seems both contradictory and (at least potentially) politically dangerous. If politics is, within itself, tyrannical and the state is a political totality, then if there is to be any justice, there has to be some way in which justice can be informed by ethics as such. But how can it be so informed without individuals engaging in a publicly expressed politics that is itself ethically motivated (and perhaps even publicly recognized as being ethically motivated), that is to say, without engaging in a politics that is not solely, or even primarily, about being concerned with a struggle for survival? As Bergo (2003, 265) asks: "How can we, as ethical incomparables who have [supposedly] become politically comparable because of the Third Party, interrupt the sway of political power?" And even if we accept that "'the concern for justice' arises spontaneously" (264), through the presence of the third party, isn't the question of *what constitutes and counts as justice* still, itself, a political issue in the sense that it is not something immediately given in the ontology of the world (as those who seek to naturalize and fix a particular political form contend)?

In other words, although one's concern for justice may be, as Levinas claims is the case for ethics as such, beyond being—that is, it precedes and exceeds what is (ontologically) given—the question of what justice requires is not something determinable a priori or something that can be decided by one for all others (for that is, by definition, tyranny) but something that has to be approached through the in between of politics understood as the public expression of numerous *differently* human individuals, those who, if Levinas is right, are themselves always constituted through the ethical call of the Other and thus have the possibility of not being simply self-serving. Even on its own terms, then, Levinasian ethics cannot inform political justice without the initial intervention of Arendtian politics, a politics that is, to some extent, always already ethical.

This does not collapse ethics into politics, or vice versa; they are not coextensive and their relations, although often coconstitutive, remain asymmetric. Ethics as such and politics as such remain anarchic in all the senses already mentioned. This understanding will not then automatically issue, as Levinas seemingly hoped, in a politics (or a political state) in the service of ethics—"politics must be controlled by ethics" (Levinas in Simmons 1999, 92)—nor, as Critchley apparently thinks, an ethics "for the sake of politics." Nor, even admitting that justice is necessary, will it naturally result in a given political form, still less a state form, that somehow universalizes or captures the essence of this (anarchic) relation. Instead, it brings us back to an earlier thesis: *If politics as such is a matter of pure means, a means without end, then ethics as such could be thought in terms of being concerned with others as impure ends, that is, as beings of indefinable (infinite) value but finite worldly existence.* And these matters of ethical and political concern can be thought in anarchic terms: as expressions of singularity, natality, saying, and responsibility.

Earthly Associations: Ethics and Politics

> Anarchy cannot be sovereign, like an *arche*. It can only disturb the state—but in a radical way, making possible moments of negation *without any* affirmation. The state then cannot set itself up as a whole. But, on the other hand anarchy can be stated.

> —Emmanuel Levinas (1998a, 194n3)

No doubt the relation between Arendtian politics and Levinasian ethics could be explicated in other ways, but it is necessary to focus here on the ecological implications of this anarchic understanding and to do so outwith[10] Arendt's, and especially Levinas's, own humanist presuppositions. This is necessary for both an ethical and a political reason. First, Levinas's ethical thought concerning the specificity of the Other has anthropological limits, revealed through comparison with Murdoch's open ethical texture (see chapter 2) and also, for example, by the critical commentaries of Llewelyn (1991b; 1991c), Wood (2005), and Derrida (2008). As we have seen, Levinas's metaphysical presuppositions about the special quality of the encounter with the human face preclude any straightforward extension or application of his theory to an ecological ethics. Second, this same metaphysics ensures that the ghost of the anthropological machine haunts his own failure to articulate a politics that might be more than matters of self-interest, survival (bare life), and the sovereignty of the nation-state.

To even begin to understand the ecological potential of ethics and politics as such, it is thus necessary to recognize that the words and deeds of environmentalists (among others) attest that ethical concerns for our (differently) nonhuman neighbors are both possible and politically important. Of course, those beholden to a modernist constitution founded on the separation of nature and politics may refuse to countenance such a possibility, but then, as already argued, ethics and politics as such concern themselves with creating constitutive associations and are not beholden to constitutionally imposed limits on their activities.

In any case, the best that could be hoped for, given a politics that accepted Levinas's metaphysical (and perhaps, ironically, ontological— see Wood 2005) strictures on the ethical specificity of the human, is an extremely anthropocentric form of distributive environmental justice: a form of ethicopolitics that regards every nonhuman aspect of the natural world as a resource to be allocated by and for humans, hence, ultimately, reducing ecological politics to those political frames of reference that already exist. Such a politics is exemplified in recent attempts to distinguish the environmental justice movement, which Shrader-Frachette (2002, 6) defines as "the attempt to equalize the burdens of pollution, noxious development, and resource depletion," from wider ecological concerns, and to then claim that all ecopolitical questions can be answered in terms of social justice (Sandler and Pezzullo 2007).

There are many variations on this theme, whether Marxist, socialist, or liberal (see, for example, Pepper 1993; Wissenburg 1998). But even if issues of social and distributive justice are deemed important constituents of any environmental politics, most ecologically concerned political theorists still recognize that this, by itself, is not enough. For example, Hailwood (2004, 3),[11] while accepting many elements of a traditional liberal theory of justice, also claims that green politics are "non-instrumentalist, and so say that humanity and human interests are not the be-all and end-all" (see also Harvey 1996, 172–73; Barry 1999, 262). These theorists at least recognize something of the constitutive reality (if not always the constitutional validity) of ethical concerns for nonhuman nature and the political importance of expressing (and discussing) these concerns publicly. Some go so far as to suggest the need to develop a specifically "green communicative ethics/politics," referring to an "ecological democracy" (Dryzek 1990).[12] The important point is that any politics that accepted Levinas's strictures about the necessary *Humanism of the Other* (Levinas 2003) *would be more, not less, restrictive than most existing varieties of ecological politics.*

So an ecological politics has to be informed by an ethics that exceeds Levinas's humanist presuppositions, supplemented by a liberating Arendtian understanding of politics as such. And although Arendt, like Levinas, was almost entirely concerned with the *human* condition, although the "who" that appears in the political space is always (differently) human, there are no a priori restrictions on what or who these persons might be concerned to speak about or for. Although expressing that *amor mundi*—the love for the world that Arendt also felt—may have no constitutional place, there is nothing inherently apolitical about ecological concerns.

But this does suggest that we need to ask what kinds of ecological politics might express such concerns, raising questions about the relation of ethics and politics as such to (ecologically sensitive) forms of politics—about whether this "anarchy" is, as Levinas's quotation suggests, only a "negation without *any affirmation*" (whether, or in what sense, it might be nihilistic) and whether it can actually be stated both in terms of enclosing its saying within what is said and in terms of its being enclosed within and reduced to associations within current constitutional (state) forms. ~~*and his*~~

Ecological politics is actually far from being nihilistic if this is understood in the sense of a total (and totalizing) rejection of everything present, of everything that has ever been said, of every moral norm, of every aspect of all ethical and political systems—it is not just a critique of all there is, in the name of nothing at all, although it certainly initiates a continual critique of all (ethically and politically) restrictive deployments of metaphysical absolutes (in Vattimo's sense). Nor is it nihilistic in many other senses—not least because it is through such political engagement that we exercise our freedom, use our judgment, and appear as that singular individual who we are, before others. More important, and in Levinas's terms, less "egotistically," this politics is concerned with facing and sustaining ecological beings that are not congruent with, nor reducible to, my own self-possessive interests, beings that, as Murdoch remarks, often appear alien and incomprehensible. An ecological politics as such emerges through facing up to and recognizing our potentially infinite ethical responsibilities for Other (more-than-human) beings. It is difficult to think what could be less nihilistic than this, or more life affirming. In other words, Levinas's

claims his putting forward a life-affirming politic

negativity without any affirmation might be better understood in terms of Bataille's negativity with no use (see chapter 3).

This ecological politics is informed by an ethics of responsibility before and beyond all else, before even the claims of justice. For justice can be couched in terms of a compromise between the (selfish) needs of all those (human) "citizens" concerned and metaethical formulae and/ or constitutionally defined governmental practices designed to ensure that all (who count as citizens) are treated equally, processed as ethically and politically the *Same* in an abstract sense. Such a system is clearly exclusionary both of ethics as such and of those who cannot speak for themselves (at least in ways that the system could hear or accept). Leaving open the possibility of attending to the singularity of each ecological instance requires ethically informed political interventions, interventions already accepting some responsibility for more-than-human others as Others.

It is also important to recognize that even if justice is couched in terms of the apportioning of ethical responsibilities, to the extent that this could ever be applied *systematically* without the continual reappearance of politics as such to trouble its decisions, it would risk becoming dangerously biopolitical and even totalitarian. And without politics as such, even systems of ecological justice that explicitly recognize some ethical (or, perhaps more accurately, moral) obligations to the more-than-human world run this same risk. Like Westra, they forget that there is no natural justice, that justice never follows automatically from, nor is simply a matter of establishing an accordance with, the ontology of the world. For example, when Baxter (2005, 1) claims that "ecological justice . . . can (and I think should) be given a basis in some form of naturalistic ethics," he effectively reduces ethics and politics as such to an antipolitical (in the Arendtian sense) and an unethical (in the Levinasian sense) philosophical task—namely, to determine the foundational ontological criteria (first principles—*archē*) for apportioning ethical responsibilities—criteria based in what (not whom) that being is defined as being. Such forms of ecological justice, however well intentioned, fall into the trap of naturalizing (and depoliticizing) a fixed ecological order.

What is more, the understanding of ethical responsibility being developed here is anarchic and asymmetric from its inception. Both Arendt and Levinas emphasize the radical asymmetry of the associative relations among individuals: "the asymmetry of intersubjectivity"

(Levinas 1998, 105). The singular individuals of Levinasian ethics and Arendtian politics are, for example, in no way reducible to the sovereign, self-interested, and incorporeal abstraction that is Homo economicus. Consequently, neither Arendt nor Levinas (at least when we leave aside the matter of the Third) is interested in what would be (un)ethical or (a)political communities that can only exist theoretically on the premise of sameness or equivalence—such as an idea(l) of an essential, shared, or universal human nature, or based only (or at all) in reciprocally beneficial exchange. Theirs are associations constituted in irreducible difference and plurality. And it is this asymmetry that opens the very possibility of an ecological politics that can envisage ecological communities of (differently) human and more-than-human beings. Only such ethical responsibilities *and* their political expression can constitute an ecological politics as such, one that recognizes the traces left by the diverse denizens of the world and where it is possible to speak and act concerning the more-than-human world.

This affirmation is already suggestive of certain political possibilities, but it still leaves unanswered the question of whether, or to what extent, an anarchic ethics and politics as such can be, in Levinas's words, "stated'—encompassed by what is said within, for example, a particular theory or political ideology and/or a particular state form.

Savage Democracy: Ecology in the Political Wilderness?

An obvious consequence of emphasizing the ways in which ethics and politics are constitutive of our ecological and human associations is that there cannot be any a priori or definitive answer as to how to found political constitutions, institutions, and ideologies on principles *(archē)* that (claim to) guarantee political stability or particular outcomes. Neither ethics nor politics begins with, requires, or can be contained within first principles, and the myriad forms and contents they take are inseparable from their expression and enactment. Yet this does not mean that an understanding of ethics and politics as such is without consequence in the sense that those who espouse such a perspective remain neutral about either the forms taken by government/governance or the content of ethicopolitical ideologies. Clearly, such an understanding is placed in opposition to those systems of government and thought that want to manage and suppress ethics and politics as such, that are dependent on reducing human life to bare life, that seek

a priori- Those that come before hand and opporate as an end to impose a monotheistic conception of the properly human and deny plurality and diversity or to replace ethical responsibility with moral rules and/or political responsibilities with automatic obligations. It is antitotalitarian and antibiopolitical, valuing of singular expressive freedoms. In other words, it sets itself in opposition to many forms of government and ideology in their entirety. And, being against the principle of sovereignty in all its guises, it is, in its pure form, anarchist— although, as already argued, any claims to encapsulate such purity/ innocence even (or perhaps especially) in a state of nature, are actually culturally conditioned myths.[13]

This leaves open the possibility that some forms of government may be preferable to others, more amenable to sustaining (or at least less actively repressing) ethics and politics as such, more tolerant of difference and diversity, more concerned with leaving open possibilities for individual expression, and this is something only the purist would deny. But it is also to recognize that this *never* equates with a claim that ethics and politics as such are reducible to participation in the constitutional machinery of government. Indeed, ethics and politics as such most frequently take the form of extragovernmental community and social activism (and of a civil society very broadly understood) of one kind or another. The point is to recognize that constitutional questions are, as they were for Arendt, secondary questions: they come only after, and are dependent on, the enactment of ethics and politics as such (which again is not to say such questions are unimportant). This is why Arendt constantly emphasizes the natality of the political event, the totalitarian dangers inherent in replacing politics with *policy* or bureaucracy (Arendt 1975), the vital importance of exercising individual ethical and political judgments in all circumstances (1994), and the need for civil disobedience in some (1973). She even specifically criticizes the U.S. Constitution (1965, 232), which she otherwise regards as genuinely revolutionary in intent, for leaving precious little space for politics as such.

To say this is not necessarily to agree with Arendt's own political views but to give due recognition to her emphasis on politics as such and to point out that even commentators like Waldron (2000, 203), who explicitly try to emphasize the constitutional aspects of Arendt's writings, have to admit that Arendt values political events "primarily for themselves," that is, under their constitutive aspect, as "pure means." And from Arendt's perspective, a constitution's political purpose should be to sustain the open and free texture of politics and ethics, a space of plurality and of persuasion without compulsion. Any discussion of the (de)merits of governmental forms and of the degree to which ethics and politics as such can or cannot be stated (given governmental form) has to recognize this.

It is also important to remember that this anarchic articulation of the relation between ethics and politics is not necessarily ecological in and of itself, just as the understandings of Levinas and Arendt are not necessarily ecological (although Murdoch is, perhaps, a slightly different matter). It only becomes ecological when understood in relation to the twisting and weakening *(Verwindung)* critique of the plots and metaphysical presuppositions of the anthropological machine and of the claim to human sovereignty over the natural world. It is here, in displacing and dissolving the anthropological centrality of the category of the properly or fully human in both ethics and politics, that ecology expresses a truly radical break with any and all previous political *archē.* But this critique is not the sole property of radical ecology. It is clearly articulable within, and even a vital constituent of, the political writings of many of those whose primary concerns are certainly not ecological, such as Arendt and Agamben (but also Vattimo, Foucault, Derrida, Bataille, etc.). And so there is "common ground" here where discussion and persuasion can take place, and once the incessant productions of the anthropological machine are stilled, the walls that exclude ecological ethics may crumble, be pulled apart, and fall, much like the Berlin Wall.

Such possibilities arise because radical ecology is an underdetermined association of ecology, ethics, and politics because of its natality and constitutive potential, And this potential also means that it resists being stated in the sense of being encapsulated within formulaic or programmatic party manifestos (even those of a Green Party). Any such attempt can only be indicative and provisional as, for example, Arne Naess (1972, 1979, 1989; and see Smith 2001a) argues, even in relation to his own platform for the deep ecology movement and his ecological philosophy—"ecosophy T" (the *T* being specifically intended to indicate that there are many other possible ways of thinking ecologically—A, B, C, etc.). Such an understanding does not result in an ecologism (Dobson 1995; Baxter 1999), a novel ideology to rival liberalism, socialism, Marxism, or anarchism. Radical ecology's concerns and involvements exceed any attempts to enclose it within and place it under a set of first principles, and it has no interest in asserting itself as a new ethicopolitical world order (Ferry 1992), as an *archē*, or a new form of Green sovereignty. To say that radical ecology is not a specific political ideology is by no means to claim that it escapes ideology in general—as Althusser (1990)[14] famously declared, Marxism escaped ideology by aligning itself with the knowledge provided by the natural and (post-Marxist) social sciences—even if that science is, in this case, the supposedly "subversive science" of ecology. Radical ecology cannot pretend to provide a singular truth or to be value free—quite the contrary—although it too recognizes, as those favoring the term *ecologism* have also contended (Smith 1998), that ecological politics is relatively autonomous in terms of its concerns (it is certainly not reducible to any previous *ism*) and that there are "family resemblances" between its various elucidations.

While far from presenting a uniform political ideology, radical ecology is certainly not without critical and affirmative content. Critically, for example, in rejecting the claims of human sovereignty over the natural world and the subsequent and systematic reduction of that world to resource or standing reserve, radical ecology opposes the fundamental principles and global dominance of corporate capitalism. Capitalism's systematic commodification of the more-than-human world through its reduction to exchange value remains unquestioned and unquestionable because of the (originally theological) assumption of human sovereignty. Radical ecology is anticapitalist insofar as capitalism depends not only on the alienating expropriation of that "surplus" value that Marx argued was created by human labor but on the biopolitical stripping of all ethical possibilities from (differently) human relations to nature. But this also means that it is no less opposed to any Marxism that similarly reduces more-than-human singularities to just their human *use* values. Despite Marxism's radical pretensions, the core concept of use value deploys the same (anthropological) sovereign principle *(archē)* as capitalism; it is complicit in systematically and universally reducing nature to its value in serving human needs (however defined). This complicity should not be surprising, since use values and exchange values are, as Baudrillard (1992) argues, two sides of the same "productivist" coin (see also Smith 2001a and chapter 3).

But what, then, is affirmative about the attempt to sustain an anarchic, ecologically informed ethics and politics? And here, what is being asked, at least insofar as the answer expected is not a total(izing) vision of a political order deduced from first principles, is what *form* might ethics and politics take other than the constant critique of ruling principles and contemporary political structures. The assumption behind the question is that taking responsibility for the ethical and political effects of one's words and actions without some form of metaphysical and/or institutional regulation is somehow impossible, insufficient or entirely negative. And yet ethics and politics as such are practiced every day, even in the most adverse of circumstances, often without ever having been formulated as such, and without seeking or requiring the permission of some higher authority.

If arguing that ethics and politics as such pervade our social and ecological interactions is deemed insufficiently affirmative, this is because those asking this question usually require an answer that would fit their own expectations, that dismisses these mundane forms as having little or nothing to do with the founding of a new political order or espousing new political principles. The question represents the presumed necessity of securing a vision of how to found an alternative and faultless system in its totality, the seeming impossibility to escape "the belief that somehow or other such foundations . . . [are] necessary, the belief that unless there are foundations something is lost or threatened or undermined or just in question" (Mouffe 2005, 15). This, then, is a question that mirrors those posed to atheists by those convinced that the authority of God must be replaced by some other overarching principle, preferably one guaranteed by the authority of a unitary institution—a secular equivalent of the one true Church. It takes the same form as the question posed to anarchists by political realists incapable of thinking that politics might be initiated, organized, and practiced outside of the state form (and this despite its relatively recent historical origins) and who demand to know what form an anarchist state would take, and how it would guarantee its hegemony.[15] It reflects the dissatisfaction moralists have with the anarchic aspects of Murdoch's and Levinas's ethics, which again seem to offer no permanence, no solid foundations, only a constant questioning—"Is this good (enough)?"—and a finite life facing infinite responsibilities.

Of course, there might also be a pragmatic aspect to this question, concerning how best to sustain ethics and politics as such in very different circumstances. But the fact that there are very different circumstances suggests that there can be no one universal answer. To provide such an answer is the function of an ideology like liberalism, Marxism, or anarchism. But since *isms* are not options, what seems lacking here is not an alternative set of principles to replace sovereignty but ethico-political concepts capable of articulating some of the possibilities that arise from affirming the value of ethics and politics as such, of provisionally expressing provisional (not authoritative, complete or ultimate) political forms. And the danger in providing such a vocabulary is that it immediately becomes open to ideological abuse—it ceases to be treated as provisional. For this reason, these concepts must be envisioned in terms that (so far as is possible) resist the logocentrism that places words themselves in some position of absolute authority over the world, ethics, and politics, the use of language as a definitive expression of human "authority."[16] *Provisionality* must recognize, in Levinas's terms, that saying is never fully encapsulated in the said, and moreover, that what is said should not be reified as if it were itself a principle (or essential) aspect of reality. To adopt Lacan's (1992) terms, the "symbolic order" (like the "political order" and also like Hegel's historical dialectic) never captures the "real," that is, the world outwith but informing that order, no matter how intense the desire may be to attain completion, to give symbolic designations to, and conceptually grasp everything that appears to be lacking in its current constitution. There is an affinity between this Lacanian notion of the real and his employment of it to critique and weaken logocentrism and Levinas's understanding of the ethical Other, one that, as certain theorists have noted, has political implications (Stavrakakis 1999).

The real, in this Lacanian sense, is that which is lacking in (outwith) the symbolic order, but it might also be thought of as that which remains wild, ungraspable in its essence, resistant. Consequently, the symbolic order can never actually be sovereign (and Adam's task of naming the world is interminable) because every attempt to *include* an aspect of the world within the symbolic order is simultaneously an *exclusion* of the real, of that which, despite its intentions, evades its grasp and over which it can, through definition, have no authority.[17] And this is also why words can be (as Arendt argues) political and why they cannot and should not serve to *compel* agreement—they are always radically incomplete and inconclusive in terms of the real world. They offer only provisional interpretative understandings whose persuasive potential resides in how well they manage to express something of differing practical involvements (activities) in the world—that is, particular forms of life, in Wittgenstein's (1988) sense (see also Davidson and Smith 1999).

The provision of such concepts should not be thought in terms of defining a new political ontology, at least not in Levinas's understanding of ontology (see chapter 2). Rather, as Sandilands (1999, 186) argues, the "nature of politics must be spoken by humans who are cognizant of the limits of speech." And this also means that "an inevitable failure of democratic representation must be explicitly included in a radical democratic project if nature is to be represented at all; if the part of nature that is beyond language is to exert an influence on politics, there must be a political recognition of the limits of language to represent nature, which means the development of an ethical relation to the Real" (180). This, again, is why Evernden is wrong in asserting that ecological ethics and politics is only concerned with the stripping of earthly meaning for humans (see chapter 4). And although Sandilands does not refer to Levinas in her description of an ecofeminist ethics and politics, her argument seems in keeping with his understanding of ethics as a concern that reaches beyond being (ontology), beyond that which can ever be made fully present in the symbolic order of language.

Sandilands develops this innovative understanding into a provisional account of a radical ecological politics as a form of participatory democracy, "a relation between human and nonhuman in which a democratic conversation is simultaneously valorized and recognized as always already incomplete" (181), and where other (more-than-human) beings are recognized as affecting and yet escaping full representation within human language. And if this lack of full representation seems to present an insuperable barrier to expressing an ecological politics altogether, this is largely because the anthropological machine and its modernist political scions have been presaged on the view that human language can somehow fully represent individual human realities (something Levinas and Lacan explicitly deny). Such a position would ultimately suggest that even democratic politics has no need of ethics as such, since all democracy requires is to recognize what has actually been said and done, not what is (always) left unsaid or of the attempt to say something as yet unheard, not in trying to understand and interpret what might be behind these words and deeds. This would, effectively, make politics a dead letter. And far from simply dismissing the additional difficulties in speaking of and for the (differently) nonhuman, of being an advocate, in however limited a sense, for the more-than-human world, these complexities are precisely the preserve of ecological ethics as such. These difficulties, this inevitable lack in nature's political representation, are why ecological ethics can never be a political side issue.

Sandilands suggests one way (and there are many others) of provisionally conceptualizing a radical ecology, one that, to some extent, fits with and develops what Claude Lefort has referred to as "savage democracy." By this, Lefort signals the need for the continual questioning of all attempts to present a complete sociopolitical order, to attain a totalizing ideological hegemony over politics as such. His primary targets are forms of totalitarianism, "a mode of socialization based on a fantastic denial of division" (Abensour 2002, 706), where a unity of purpose is imposed on all and a particular ideology invades and strives to achieve hegemony over all aspects of life through "monopolization of the means of coercion, information, and indoctrincation," a situation, as in the former Soviet Union, where the "dimension of the Other found itself, if not abolished (how could it be?), then at least effaced" (Lefort 2007, 141). Totalitarianism, we might say, attempts to reduce politics entirely to biopolitical management in the name of a (supposedly unifying and universal) ideology. But, as Bourg (in Lefort 2007, 15–16) argues, echoing Arendt, "If everything is political, nothing is. The attempt to totalize representation leads to the destruction of representation. The symbolic negation of otherness and difference is related to the literal negation of others identified as different and abnormal outsiders."

Lefort, whose later work is also indebted to Arendt's understanding of politics, recognizes that these (biopolitical) tendencies are by no means limited to totalitarian regimes. In this sense, the notion of savage democracy is intended to provide an alternative (and provisional) concept that highlights both the importance of politics as such and the difficulties in political (rather than a more general symbolic) representation. And while Lefort only occasionally refers to Lacan and is not generally dependent on using Lacanian notions of the real (see Flynn 2005, 119),[18] this connection between symbolic and democratic representation is further developed by Sandilands, following Stavrakakis's (1999, 73) argument that if we attend to the "political per se," what "constantly emerges in . . . contemporary political theory is that the political seems to acquire a position parallel to that of the Lacanian real. . . . The political becomes one of the forms in which one encounters the real."

This connection seems fruitful because for Lefort (and Sandilands) there is a "positive [affirmative] aporia" (Abensour 2002, 707) at the heart of democratic politics that also entails a continual critique of the attempt to define and locate democracy. (The question of what constitutes democracy is itself a vital aspect of savage democracy, something never agreed upon and constantly discussed.) Democracy is not a matter of defining what is or should fill this empty place, that is, what politically "reality" actually is or should become, or how it should be captured by or within the symbolic structure of a regime, precisely because democracy both requires (and requires the recognition of) the existence of this aporia, the impossibility of completing politics as such—"democracy requires that the site of power remain empty" (Lefort 2007, 143). As Sandilands (1999, 187) puts it, "In the absence of the embodiment of power, all law and knowledge circulate around the empty place; nobody has 'it' right, nobody has the right to claim the foundational power of the political order, ideological desires to the contrary notwithstanding."[19]

And this indeterminacy might also be interpreted as a claim that (savage) democracy is incompatible with the principle of sovereignty, something Lefort explicitly recognizes with respect to the change from medieval conceptions of the figure of the sovereign as the actual, literal, embodiment of authority to modern democratic forms. For this change to occur, "it is necessary," says Lefort (2007, 43), "that the sovereign have ceased to incarnate the community and that he no longer appear above the law." Democracy requires the "disincorporation of power" (143). Lefort (unlike Agamben) is, however, less clear about how the principle of sovereignty continues to operate in those actually existing modern societies that make this claim to be democratic, or to what degree current state forms of democracy can and do embody a notion of savage democracy. And so, while recognizing that "anthropological evidence" supports the primacy of (what has been referred to here) as constitutive power,[20] and arguing that "sociology on the one hand and phenomenology on the other unveil for us a network of originary relationships; more than that, there is an intricate connection of beings, perceiving, thinking, and acting in their common world that underlies the symbolic constitution of every community" (2000, 155), Lefort still seems to think that it is "futile to deny the notion, everywhere present, of a pole of sovereignty" (155). This suggests that Lefort does not recognize the vitally important difference between a Schmittian definition of sovereignty as exception and ruling principle *(archē)* and the pervasive presence of more general, symbolically mediated representations of political authority.

Despite this serious theoretical deficiency, the notion of savage democracy resonates well with a focus on politics as such. For example, Abensour (2002, 707) argues that in using this term Lefort, "intends to dismiss the claim to reduce democracy to an institutional formula, to a political regime, or a set of procedures or rules," that is, we might say, to a constitution, emphasizing instead that a democratic understanding of politics as such requires a "refusal to submit to the established order" (707). Interestingly, Abensour also describes Lefort's notion of savage democracy as "a movement of indetermination" (708), as evoking the spontaneity of the "wildcat strike" that "begins within itself and unfolds in an 'anarchic' fashion, independent of any principle *(archē)* or any authority" (707), and as a refusal of domestication—"democracy is not domestic or capable of being domesticated to the very extent that it remains faithful to its 'savage essence'" (707–8).

Abensour certainly exaggerates the anarchic and anarchist aspects of Lefort's thought (of which Flynn [2005], by contrast, is excessively dismissive),[21] but savage democracy still offers possibilities for provisionally rearticulating politics as such. Of course, the ecological ramifications of this move would depend on developing a less restricted understanding of *savage,* one that calls to mind the etymology of savage not only in terms of being resistant to domestication, of wildness, but of relating to the more-than-human wilderness and wooded *(sauvage/* sylvan) worlds, relations that Sandilands (1999, 194), too, terms "wild." But here again, the anthropological limits to (in this case Lefort's) modernist thought come to the fore because the empty space of democracy is perhaps not as symbolically empty as it might seem. Indeed, this space and the mantle of sovereignty is occupied by the symbol of "the people" *(demos)* who, while recognized as a plurality, are always and only (albeit differently) human.

Lefort (1986, 303–4), quite rightly, focuses on the dangers of slipping from a notion of democracy as "ungraspable, uncontrollable society in which the people will be said to be sovereign, of course, but whose identity will constantly be open to question, whose identity will remain latent" to one ruled by the protototalitarian fantasy of "the People-as-One, the idea of society as such, bearing the knowledge of itself, transparent to itself and homogeneous, the idea of mass opinion, sovereign and normative, the idea of the tutelary state" (305). This fantasy of completion is a constant totalitarian danger for any democratic polity, and he is certainly aware that the "bourgeois cult of order which is sustained by the affirmation of authority, in its many forms, by the declaration of the rules and the proper distances between those who occupy the position of master, owner, cultivated man, civilized man, normal man, adult and those who are placed in the position of the *other* . . . testifies to a certain vertigo in the face of the void created by an indeterminate society" (304). There is, one might say, a "fear of political and individual freedom," a fear that has been connected to the rise of totalitarianism by thinkers as diverse as Erich Fromm (2001) and Carlo Levi (2008), an insecurity born of an unwillingness to face the provisionality of, and take responsibility for, both our human self-identities and our political forms. However, Lefort fails to recognize that this danger to pluralism and such declarations of "proper distances" are present in the principle of sovereignty itself and the mode of operation of the anthropological machine; that is, the principle, even of *democratic* sovereignty, operates by and through decisions that enclose and claim to fill this empty space (envisaged as a power vacuum) by and through the inclusion/exclusion of those who it is decided do not properly constitute "the people."

The principle of sovereignty, even the sovereignty of the people, and especially the sovereignty of the people over the Earth, has to be rejected; otherwise we are in constant danger of replacing Arendt's political spaces of encounter—and of concern, discourse, persuasion, interpretation, and responsibility for others—with a symbol that reifies the people as an abstract ruling (constitutional) principle rather than an actual living (constitutive) plurality. And, as argued previously, this is precisely how the anthropological machine works.

Many theories of radical democracy do go some way toward recognizing the dangers of operating on the basis of an abstract principle of humanity. For example, Chantel Mouffe (2005, 13) argues, "Radical democracy demands that we acknowledge difference—the particular, the multiple, the heterogeneous—in effect, everything that had been excluded by the concept of Man in the abstract." And other recent theorists, such as Paolo Virno (2004) and Michael Hardt and Antonio Negri (2004), have tried to respond to such concerns by developing new provisional concepts like that of the "multitude." The multitude is a form of political "organization [that] emerges through the collaboration of singular subjects . . . this production of the common is neither directed by some central point of command and intelligence nor is it the result of spontaneous harmony among individuals, but rather it emerges in the space *between,* in the social space of communication. The multitude is created in collaborative social interactions" (Hardt and Negri 2004, 222).

Although this understanding places more emphasis on collaborative social interactions than Arendt's notion of politics would assume, it is nonetheless a description of a constitutive (or in Hardt and Negri's terms, "constituent") form of politics as such. Here too, then, we might find some elements of a provisional language to express what is affirmative in a radical ecological politics. However, like Lefort, these thinkers are caught up in anthropological machinations. For example, Mouffe's critique of the abstract concept of Man is by no means linked to a critique of human sovereignty, and the differences she recognizes as ethically and politically important are not those associated with the more-than-human world. Even Hardt and Negri, who are much more openly critical of the principle of sovereignty than either Lefort or Mouffe, and who focus explicitly on its biopolitical ramifications, still assume this same sovereign principle in its anthropological form.[22] Thus, although they recognize the political importance of environmental activism (Hardt and Negri 2004, 282–83), these actions are automatically translated into questions concerning human use values. For all their conceptual innovations, Hardt and Negri remain entirely within and dependent on Marxist understandings of the relation between use values and exchange values. "In the anthropological historicity and substantial temporality characterizing the transformation of ways of life, use value is always recovered as a basic element. Although it is continually modified, it always remains fundamental to the revolutionary project" (Negri 2008, 83). But to deem use values the fundamental political principle is already to be engaged in biopolitics, to reduce nature to standing reserve and politics, foremost, to matters of survival, that is, in Arendt's terms, to make politics subservient to, and ultimately judged by, categories appropriate to labor. *"Use value is living labour"* (Marx in Negri 2008, 85).

There are then very serious problems with Hardt's and Negri's approach from the perspective of any radical ecology, as there are from that of Arendt (whom Negri [2008, 160] misreads as being primarily concerned with formal constitutional politics) and Agamben (whom Negri [88] claims is too indebted to anarchic understandings of ethics and politics). Negri is also very skeptical about any ethics of responsibility (172), although he reads this too in a rather narrow way. Nevertheless, the idea of the multitude has certain provisional possibilities insofar as it emphasizes both the singularity of those who compose the field of politics *and* its constitutive power. (And there are many other ways of thinking about a postfoundational and provisional politics, for example, through the "left-Heideggerianism" of Jean-Luc Nancy's [1991] "Inoperative Community," discussed later, and see Smith 2010.) What is clear, though, whether using either Lefort's concept of savage democracy or Hardt's and Negri's multitude to refer to a radical ecological politics is that this has to be taken as provisional in the sense both of being open to change and sustaining its own development through the provision of the spaces necessary to *be* (to exist) differently. It has to be both ethical and ecological because such a polity exists only to the extent that people are willing to take responsibility for their words and deeds—which means responsibility for their effects on Others, including more-than-human Others. These kinds of concerns have long been articulated by various forms of radical green politics, which emphasize participatory decentralized democracy, nonreductive understandings of bioregionalism, the importance of thinking globally while *acting* locally, and of treading lightly on the Earth.[23]

This ecopolitics is a politics of singular freedom and responsibility that is both antitotalitarian and antibiopolitical. It persists only where nature, no less than humanity, escapes being entirely managed and controlled and for as long as *diversity* (human and more-than-human) is expressed and sustained (it is never monocultural). It emerges first through face-to-face encounters (by no means limited to Other humans) and is therefore born of and sustained by relations of proximity that are both intimate and embodied, relations that can be expressed (if not defined) through our subsequent words and deeds. It recognizes the finitude (mortality) of Earthly existence, the unique and irreplaceable singularity of every life that touches us and is lost, and the infinite responsibility that dispossesses us, a responsibility we would face alone were it not for the potential forgiveness of our fellows. It is not based on any natural(ized) laws nor beholden to any overarching universal principles. It resists the way that capital (and its lack) constantly threatens to turn people and places into shadows of themselves.

But beyond all this, an ecological ethics and politics of responsibility troubles and contests the key principle and supposed origin of all modern constitutional politics—the self-possessive principle of sovereignty, whether the sovereignty of the individual self, the word, God, the Good, the nation-state, the people, or human sovereignty over the natural world (Smith 2008a). It does so because this often taken for granted and naturalized principle abrogates the constitutive power of others to itself and grounds its own powers in its ultimate ability to strip Others of their ethical and political possibilities, even of their very lives, and because ultimately it harbors the potential to turn the whole world into a realm of crisis and disaster.

7 AGAINST ECOLOGICAL SOVEREIGNTY

Globalization, Modernization, and Gaia

[handwritten margin note: For smith, sovereignty is inherenty, anti-green, Smith doesn't Think can green The state]

Several recent texts (most notably, Eckersley 2004) have argued that given the ecologically destructive effects of unfettered economic globalization, good, pragmatic arguments exist for environmentalists to advocate and support a form of green state sovereignty. However, the key question concerning this strategy is not necessarily the plausibility, or as Paterson (2000, 45) argues, the implausibility, of "greening" state institutions but the biopolitical dangers to ecology and politics that sovereignty itself represents.

The potential erosion of state sovereignty has become a frequent theme in discussions of ecological aspects of globalization. And while there are many different understandings of globalization in terms, for example, of the time–space compression brought about by faster and more extensive transport and media technologies (Harvey 1996, 243; Schivelbusch 1986), increasing economic and sociocultural interdependencies (Robertson 1992), and the proliferation of a wide variety of supra- and nongovernmental global organizations and agreements, the primary driving force behind such tendencies is global capitalism. Even influential theorists like Manuel Castells, who emphasize the transformative power of global information technologies in the emergence of a new transnational "network society," recognize that such technology is itself "shaped, in its development and manifestations, by the logic and interests of advanced capitalism, without being reducible to the expression of such interests" (Castells 1998, 13). The statism of the former Soviet Union and its attempts at social restructuring *(perestroika)* collapsed, Castells argues, because it failed to adjust sufficiently to the globalizing forces of "informationalism" and its associated form of capitalist restructuring.

Globalization, then, especially economic globalization, is frequently portrayed by proponents and detractors alike as an almost irresistible process (a form of historicism) that is inimical to the retention of traditional state sovereignty. To (mis)appropriate Bill Clinton's remark, "It's the (global capitalist) economy, stupid," and political luminaries and plain citizens of every nation-state are expected to understand that their continued well-being is entirely dependent on adapting to the constantly changing circumstances that this system generates. The political uncertainties of a post-Westphalian world, one where states are not even theoretically free and equal to enjoy "ultimate authority over all objects and subjects within a prescribed territory" (Benhabib 2006, 23),[1] are conjoined with the social uncertainties that characterize the "forms of life which are emerging as the global begins to replace the nation-state as the decisive framework for social life. This is a framework in which global *flows*—in mediascapes, ethnoscapes, finanscapes, and technoscapes—are coming to assume as much, or greater, centrality than *national institutions*" (Featherstone, Lash, and Robertson 1995, 2). We inhabit a world of "global complexity" (Urry 2004), which transforms the roles of nation-states.

Leaving aside, for the moment, the question of whether such accounts tell the whole story, since as Litfin (1998, 2) notes, the "state is unlikely to be placed on the endangered species list anytime soon," it is fair to say that ecology, until recently a political side issue, now appears as a significant contributor to this process in various ways. For example, the pressing need to solve transboundary environmental problems caused by the prioritizing of economic growth—problems like ozone depletion, acid rain, and global warming—have engendered a series of international agreements (including the Montreal and Kyoto protocols), which, however (un)successfully, endeavor to place some negotiated limits on the exercise of nations' absolute territorial autonomy. Both ecological effects and responses to them thus permeate states' territorial borders.

From an optimistic point of view, such agreements are only the tip of an ecological iceberg; they exemplify just one kind of sociopolitical response to human-induced environmental changes, the cumulative effects of which, as they ramify throughout social and economic systems, are regarded as constituting a global form of "ecological modernization" (Mol 2003). In other words, the ecological repercussions of our economically driven activities are, ironically, deemed to operate, via their effects on a variety of social "actors" (including environmental nongovernmental organizations and transnational corporations), as a form of negative feedback loop, one that serves both to regulate and reduce further ecological damage and promote new and more appropriate forms of environmental governance. This, in turn, will facilitate further, supposedly more sustainable, economic growth. (In many ways, this might be seen as an optimistic neoliberal version of Beck's risk society hypothesis (see chapter 5). If we understand "ecological modernization as environmental restructuring" (Mol 2003, 61), this suggests that "roles and responsibilities formerly reserved for nation-state actors are fulfilled by market actors and civil society groups and vice versa" (Spaargaren, Mol, and Buttel 2006, 15). Environmental "politics" becomes enacted through a proliferation of diverse, ecologically enlightened, hybrid entities (such as socially responsible corporations) that combine organizational and regulatory aspects of state, market, and civil society. Worryingly, given the biopolitical associations already noted, the favored concepts for understanding this emerging situation are those of (cybernetic) self-governing systems, transboundary environmental flows, networks, and complexity.

These theories might then be regarded as the social system equivalents of James Lovelock's (1987) Gaia hypothesis, which famously regards the entire planet as a self-regulating, entropy-defying (and hence, in Lovelock's terms, "living") system. In this sense, Gaia theory posits a fundamental biophysical form of ecological globalization, stressing the evolutionary interconnectedness and interdependence of those natural systems which provide the basic infrastructure for all social systems and, indeed, all earthly life. The ecological dimensions of this cybernetic naturalism provide a suggestive explanatory framework for globalization's transnational repercussions. However, the key difference between Gaia theory and the social theoretical approaches already mentioned is that Lovelock is much more *pessimistic* about the future trajectory of human progress—regarding contemporary social systems as anything but self-regulating and as seriously disturbing the Earth's own homeostatic functions.

Although Lovelock's naturalistic model promotes a holistic ecological worldview that seems directly relevant to issues of national sovereignty, its wider political implications usually remain almost entirely unexamined.[2] Interestingly, Lovelock (2006, 13) has no faith in the efficacy of international agreements like Kyoto or bodies like the G8. In order to save the world from the immediate threat of global warming, he advocates high-tech solutions supplied by a "small permanent group of strategists" (153) whose legitimacy is apparently conferred by the international scientific community. Despite the obviously global remit of the technical solutions he considers (for example, giant space-mounted sunshades to deflect incoming light, a massive expansion of nuclear power, and so on), he seems to think that such solutions can and will be implemented only if a single "technically advanced nation wakes up to its responsibility" (151).

Superficially at least, despite his cybernetic holism, Lovelock's environmental pessimism and his skepticism about international politics seem contrary to the claims of ecological modernizers. He also seems to confer political authority on at least one, unspecified, nation, which would claim to act on others' behalf as a technologically superior and morally justified (responsible) world leader. However, this position is, if anything, even more corrosive to the Westphalian model of national sovereignty because even though it accords one state the ability to act alone, unilaterally taking such *extra*territorial powers would clearly infringe on the sovereign territorial authority claimed by every other nation. Again, the problem for national sovereignty seems to be the inescapable territorial permeability of causes and effects in an ecologically interconnected world. The apparent issue here, then, is not the Westphalian model per se, since all these analyses seemingly point to the erosion of *territorially limited* sovereignty (or at least its being radically transformed), but rather of the modes in which this erosion occurs and the degree of optimism espoused about this situation. Ecological modernizers regard globalization and its ecological effects as ushering in more or less novel forms of governance, while Gaia theorists seem relatively unconcerned about the political dimensions of their proposals or, in Lovelock's case, are even hostile to what they regard as further political prevarication.

The implications of these apparently disparate responses to ecological aspects of globalization are worth further reflection because, while to some extent they all seem to challenge the political sovereignty of those nation-states regarded as the essential elements of modern "international" relations (Kuehls 1996, 26; Paterson 2000), they also offer different accounts of the fate of ecological politics in both its broader and its more radical senses. The unilateral global extraterritoriality of Lovelock's emergency state responsiveness would not only undermine other states' political sovereignty but also, and more importantly, effectively suspend the space for politics altogether both within and beyond the state's boundaries. Here, "taking responsibility" for the management of this *state of global ecological emergency* means holding in abeyance the ethical and political concerns of all those excluded from participation in decisions upon which their survival supposedly depends. In this sense, taking global responsibility is really a morally charged euphemism for arbitrarily extending the institutional sovereignty conferred by a technologically advanced nation on a self-proclaimed scientific elite in a way that takes political power away from all others. Lovelock effectively offers a new, if confused, ecological and technocratic rationale for global sovereignty, one in which an elite operating within a single state awards itself the role of steward of humanity and of the world, acting, of course, for "everyone's" benefit, under the auspices of a new ecological/cybernetic godlike principle—under Gaia's incontrovertible natural (and hence supposedly apolitical) laws.[3]

Ecological modernization, on the other hand, regards the responsiveness of social systems as a source for governmental optimism, making a state of emergency (ecologically) unnecessary. Here, though, the normal (nonemergency) state is similarly depoliticized insofar as it becomes reenvisaged as a managerial regulator, a neutral cybernetic governor (another invisible hand) whose role is moderating certain ecological excesses while simultaneously acquiescing to the structural demands of, to use Castells's terms, informationalism and global capital. To this degree, what we might term the "normatively compliant state" loses some aspects of its political autonomy even as it takes on the role of managing and containing broader, and especially dissenting, ethical and political views—that is, as it becomes increasingly biopolitical.

It is no accident that ecological modernization's proponents remain largely silent about what happens when dissenting ethical or political views (or when ecological repercussions) pose serious challenges to the economic/informational system per se. As Eckersley (2004, 74) notes, ecological modernization "can masquerade worldwide as an ideology-free zone only for so long as it can succeed in naturalizing the modernization process to protect it from deeper questioning and social unrest." By defining their theories, and the current socioeconomic system, in terms of political *moderation* (in both the sense of "governance" and being "nonextreme") they avoid concerning themselves with how and when the survival of this system might have recourse to, and be dependent on, the sovereign powers exerted during crises or states of emergency. The elision of the fact that the current system is far from moderate—it continually generates chaotic and extreme social and ecological consequences—illustrates the similarities between ecological modernization and the doctrines of that earlier fictional resident of Westphalia, Dr. Pangloss, namely, that despite all appearances to the contrary, all will (eventually) be for the best in "this best of all possible worlds" (Voltaire 1957, 20).

Arguably, then, these apparently disparate responses and their understandings of politics and sovereignty are related in more profound ways to globalization and each other than seems to be the case—and this despite that globalization and its ecological effects are the explicit objects of their studies! Lovelock's autocratic/technocratic take on Gaia and ecological modernization's hybrid forms might be regarded as opposite sides of the same (global capitalist) coin, each acquiring ideational currency precisely to the extent that they are adjusted to the globalizing forces of informationalism and capitalist restructuring. The advantage of both the cybernetic managerialism of ecological modernization and of Lovelock's exceedingly dangerous emergency solutions is that they promise to sustain this same economic system while governing or, in Lovelock's case, abolishing entirely the scope for ethicopolitical dissent. That is, in Agamben's (and Benjamin's) terms, they might be regarded as relating to each other and to the principle of sovereignty as biopolitical norm and exception (see earlier discussion). This understanding throws a very different light on current arguments concerning green state sovereignty

Green State Sovereignty

To the extent that the modern nation-state has been the key focus (and to a much lesser extent key locus) of political debate, the pervasive influence of globalization seems to usher in a new age of political as well as ecological uncertainty. In response to this uncertainty, and being faced with the unpalatable alternatives of either the globalizing, coevolutionary optimism of ecological modernization or the unilateral state of emergency responsiveness that Lovelock's pessimism justifies, those concerned with retaining a space for ecological politics have a seemingly obvious solution open to them: the defense of a revitalized but ecologically informed notion of political state sovereignty. And this is precisely the tack recently taken by Eckersley (2004) in *The Green State*.

Eckersley, among others, argues for the gradual evolutionary greening of notions of state sovereignty in order to "reinstate the state" against the pressures of global capital "as a facilitator of progressive environmental change rather than environmental destruction" (Barry and Eckersley 2005, x). Although analyses of what a green state might look like and how effective different strategies for achieving such an end might be vary, the general picture that emerges from these discussions is of forms of green state welfarism (Meadowcroft 2005; Christoff 2005) and regulatory interventionism. "The move towards a more sustainable society demands," says Eckersley, "the regulation and in some cases proscription of a wide range of environmentally and socially damaging activities," and the state's capacity to achieve this regulation "arises precisely because it enjoys a (virtual) monopoly of the means of legitimate coercion and is therefore the final adjudicator and guarantor of positive law. In short the appeal of the state is that it stands as the most overarching source of authority within modern, plural societies" (Eckersley in Barry and Eckersley 2005, 172). However, while appealing to this source of authority might have a strategic logic, and while there is some slight evidence that certain states have begun to address the ecological welfare of at least some elements of their human populations, it is also fraught with political, ethical, and ecological dangers.

The pivotal historical role that nation-states have played in exploiting the global environment suggests that Eckersley's, Barry's, and others' arguments rely on an overly optimistic assessment of the possibilities of altering the trajectory of compliant states—especially given current trends away from welfarism in general. Of course, supporters of green sovereignty might argue that such struggle may be hard but is nonetheless worthwhile, since getting the state and its coercive powers on environmentalists' side would apparently have so many advantages. But even accepting such an argument, a question must surely remain as to what extent any normatively compliant state can or would oppose the interests of global capital, especially when, as Eckersley admits, they actually provide the "basic stability, contractual certainty, and protection of private property rights" (Barry and Eckersley 2005, 172) vital for capitalism to survive. But this too is supposed to illustrate the necessity of an appeal to a greened state sovereignty that can be employed as an authoritative trump card. What is more, since state sovereignty is assumed to be the very foundation on which international (a term usually deployed without question, as if it were a synonym for extraterritorial) collaboration, including any supranational political authorities, are built, and since such solutions initially require (at least on the Westphalian model of sovereignty) individual nations' agreement, Eckersley (2004, 239) can argue that to abandon attempts to green the state is akin to a council of despair (91–92), since it would also relinquish the political power of the international sphere by default.

This, however, exaggerates the political role of nation-states at the global level and underplays the importance of social movements, nongovernmental organizations, ethicopolitical resistance, and the potential for (and actual existence of) novel decentralized political forms linking local and global concerns. In Kuehls's (1996) terms, it falls back into standard models of state-centered international and transnational politics based ultimately on fixed, geographically defined sovereign territories rather than recognizing and exploring the possibilities of what he terms "transversal" (50) political forms. It might also be argued that the very claim that there is a need for a mechanism of global governance to regulate capital is also, as Paterson (2000, 157) perceptively remarks, "premised on the continued existence of globalizing capitalism, a continued existence that [at least some] Greens want to prevent." For this reason, Paterson, like Kuehls, emphasizes the *political* possibilities usually excluded from consideration in discussions of global environmental issues precisely because they assume that nation-states and global capitalism are the necessary givens and continuing and unchangeable realities that all other considerations must work around. And these are the shared assumptions not only of ecological modernizers and of Lovelock but of proponents of green sovereignty, too.

More important, though, Eckersley's arguments are not only dependent on an optimistic notion of the state as a politically malleable entity capable of being transformed to facilitate what is in effect a strong form of ecological modernization; they also depend on a protean view of state sovereignty. Eckersley (2004, 203) is explicit about this. The "ordering principle of sovereignty is a changing, derivative principle the meaning of which arises from the changing constitutive discourses that underpin it. Accordingly, to the extent to which the constitutive discourses of sovereignty have begun to absorb ecological arguments it becomes possible to talk about the concomitant 'greening of sovereignty.'"

But the trouble is that while sovereignty is a concept with a history, as Agamben explicitly illustrates (see also Bartelson 1996), it is by no means as malleable as Eckersley suggests. Its lack of malleability is even more obvious when, as Eckersley's own remarks suggest, the particular concept of sovereignty she seeks to apply is that which legitimates the monopoly of territorially defined coercive force by the nation-state—an understanding that developed even as such states themselves began to emerge in the sixteenth and seventeenth centuries. As de Jouvenal (1957, 169) notes, "Absolute sovereignty is a modern idea." And, as he also remarks, while much political effort has been directed to dividing and delimiting these sovereign powers, "it is the idea itself which is dangerous" (198).

It is dangerous for an ecological politics to appeal to the principle of (state) sovereignty because, as Agamben argues, it is ultimately based in a political decision the results of which are antipolitical. The principle of sovereignty is not what constitutes and facilitates political action but the constitutional limit that those wielding sovereign political power decide to place on politics as such. Sovereignty is, as this whole book has argued, deeply implicated, indeed decisive, in both the biopolitical ordering and management of populations and in the declaration of the kinds of states of emergency and antipolitical technocratic solutions that Lovelock and others envisage. The biopolitical management of populations *is* the normalization of the state of exception, the treatment of people as bare life. In other words, far from providing an alternative to either of these positions, it is the exercise of a principle of sovereignty that underpins them. And when does the nation-state paradigmatically exercise its sovereignty? When it deems its own security is threatened.

Eckersley hopes that states might exercise their political sovereignty as they are made aware of the fact that their *ecological security* is threatened by runaway global capitalism—rather than, as at present, thinking that their *economic security* is dependent on continued economic growth. This profound change is to be brought about by political and discursive challenges to the current role of the state (Eckersley 2004, 64), that is, somewhat ironically, by the influence of politics as such. And certainly, only those who think ethics and politics as such are somehow less important than, or perhaps possibilities entirely based in, the successful hegemony of a given economic form (such as free-market capitalism or certain economically reductive Marxist understandings of communism) would assume that the economy should not be under the influence of, but should be constituted along paths suggested by, ethics, politics, and ecology. But, contra Eckersley, this *does not* require the exercise of a principle of sovereignty (even a principle of political sovereignty over the economy)—it requires, as Eckersley effectively admits, politics as such in the sense of a political critique of that normative depoliticization of the economy that forms the very basis of neoliberal capitalism *and* the (normatively compliant) nation-state.

It is difficult to deny that national sovereignty has facilitated and legitimized economic globalization on the basis of decisions made by the governments of nation-states to treat capitalism as if it were, or should be, a politics-free zone—just a matter of consumer choice and the operation of the invisible hand of a (politics-)free market. Homo economicus is, after all, the combined product of the depoliticizing and defining movements of sovereignty (Foucault 2008) and the anthropological machine. We should recall that Agamben's account of sovereignty is *not* tied to a (totalitarian) notion of complete state control of absolutely everything within a given territory; it refers to the power to decide to suspend ethical and political norms in a state of exception—to the political act (the inclusive exclusion) that depoliticizes a population, reducing them to bare life through an increasingly all-pervasive (as exceptions become the rule) biopolitics. We can go further here and argue, as Foucault (2008, 292) notes, that the idea of sovereignty within modern capitalist states has always revealed "an essential, fundamental, and major incapacity of the sovereign, that is to say, an inability to master the totality of the economic field." In capitalism the economic field is that which sovereignty actually has least power over in respect of possessing "an absolute decision-making power" (293). It is this lack of being able to oversee all that occurs in every detail that leads to the development of a governmental rationality that operates by assuming a model of Homo economicus, that is, the populace as unethical, depoliticized, rational subjects seeking to maximize their own economic benefits.

Speaking in terms of globalization, we might say that the normatively compliant state takes upon itself both normal/normative and exceptional tasks but does so in ways that are, for the most part, complicit with the globalizing tendencies of capital and informationalism. Its normal/normative role is, as ecological modernizers assert, reenvisaged as a managerial node in the cybernetic governance of populations and networks. However, since this is not usually a matter that impinges directly on political sovereignty *in Agamben's sense,* the state's compliance can often take the form of various organizational hybrids. Powers are ceded to systems that are deemed to run efficiently only if they are free of political interference. Efficiency (especially economic efficiency) thus becomes one of the norms (ideals) that are used to govern and replace politics, an everyday (postpolitical) way to reduce politics and ethics to biopolitics, to economic dependency and (anti)social responsiveness rather than political in(ter)dependence and ethical responsibility (Smith 2006).[4]

What little politics remains is then reenvisioned as a tool of efficient governance, as feeding back information through prearranged channels to facilitate fine-scaled systemic adjustments to globalization's complex flows. But as Arendt argues, thinking about politics as if it were a process is a fatal mistake. It tries to turn it from an unpredictably creative activity constitutive of human communities into a systematically inhuman and dehumanizing predictability intended to supplant politics. It eradicates the real politics (the public actions and expressions based on individual judgments—and the responsibilities that are coeval with this), leaving only what Foucault (1994, 221) refers to as "governmentality."

Alongside this increasingly "normal" situation, the state still retains its "exceptional" sovereign power to decide on a state of emergency. If cybernetic managerialism reflects the apolitical optimism of ecological modernization *(there is no ecological crisis),* then the state of emergency as the introduction of a "survival footing" reflects Lovelock's pessimistic state unilateralism. The ecologically motivated incursions he posits, the giant sunshade in space, the massive expansion of nuclear power, and so on, are in this sense the environmental equivalents of the supposedly humane militarism of those technically advanced nations who, to paraphrase Lovelock (2006, 151), apparently woke up to their responsibility to wage the war on global terrorism. As already argued, if Agamben is right, the ecological result would be the same, that is, the decisionistic suppression of political liberties in the name of survival. Interestingly, Lovelock (2006, 153), against the advice of other Gaians, explicitly uses militaristic metaphors, speaking of a war against our

"Earthly enemy," and the possible need for "rationing," "restrictions," a "call to service," and "our" suffering "for a while a loss of freedom."

From this perspective, *the state's recognition of ecological crises will not lead it to encourage an ecological politics.* Quite the contrary, it will be coextensive with the imposition of emergency measures and potentially disastrous technological, even militaristic, "remedies." As Agamben argues, it is always indicative of the retention of sovereign power that the state maintains its monopoly on the legal use of force (and even violence), justified as a response to exceptional circumstances (even including the ability to decide what does or does not count as violence, as for example, in labeling many forms of environmental activism eco*terrorism*). As always, those who suffer most from the situation will, ironically, be those most likely to find themselves reduced to bare life.

Agamben's point is Benjamin's, that the history of the oppressed teaches us the political danger of states of exception becoming the norm. But in the case of the compliant state, where even the norm is a depoliticized cybernetic managerialism, it is so much easier (though paradoxically much less necessary) to explicitly invoke such a state (or to put it a different way, the more mundane such exceptions appear when they are invoked). The more modern life becomes postpolitical—constituted by supposedly apolitical norms like efficiency (rather than the constituting activity of the political community)—the more likely any non-state-orchestrated political expression is to be deemed exceptional. Both the normal and the exceptional situation involve the reduction of political life to bare life—to survival—one by gradual erosion, the other by sudden diktat. These elements combine when the state comes to use the exception (the ecological crisis) as its justification for its normal biopolitical management of populations. What is more, if the exception becomes both universal and permanent (as the global war on terrorism threatens to be and as an ecological crisis might easily become), then the compliant state-centered response is not ethico-political cosmopolitanism but the concerted global suppression of ethics and politics in order that global capitalism and informationalism might survive—hardly a happy prospect.

Whether such an analysis seems overly cynical or merely an expression of a healthy political skepticism depends on one's political situation. It certainly has Orwellian overtones, but like Orwell's dystopic vision, it is neither groundless nor without political purpose. The

almost total lack of public involvement or influence in the recent decisions of a number of states to restart and reinvest heavily in nuclear
power, Lovelock's favored solution to the emergency of global warming, might give pause for thought. The rationale here is that this is
the only (technical) solution to an ecological problem that can fit with
capital's ever-increasing energy demands—something Lovelock, who
is in many ecological senses no fan of global capitalism, *explicitly* recognizes. That there might be a different range of political solutions of
more or less radical nature is moot, since normatively compliant states
will, without extensive political opposition, risk another ecological and
social disaster like Chernobyl, the proliferation of nuclear weapons,
and so on, rather than challenge the global logic of capital.

Despite evidence to the contrary, Eckersley wants to regard this complicity between the nation-state, the claims of sovereignty, and global
capitalism as simply a matter of historical contingency. Downplaying
the question of their entangled origins (in which the very claims of sovereignty, the state, and the anthropological machine *lie*), she identifies
the problem as a progressive ceding of political sovereignty to the market rather than as exemplifying the mode of operation of sovereignty
itself as the power to decide political inclusions/exclusions.

But if sovereignty is, as Eckersley claims, really an entirely con-
tingent matter (if it is protean) and if, despite this, she still simultaneously envisages it as a principle of political authority (rather than
thinking politics in terms of its constitutive powers), we would still
have to inquire (as those promoting global free-market capitalism
surely would) what basis this claim to political sovereignty is supposed
to have. Where is this transformed principle *(archē)* of sovereignty
supposed to originate? Why is it presumed to be primarily associated
with the *nation*-state—the (biopolitically) predefined territory of one's
accidental birth *(nascence)* or of one's later naturalization, that is to
say, with that body constitutionally recognized as "*the* people"—those
that matter politically—as opposed to the natality (in the Arendtian
sense) of ethicopolitical action that composes myriad intersecting constitutive communities. And we would still have to ask, as Schmitt and
Agamben do, but Eckersley does not, what kind of action exemplifies
the exercise of sovereignty?[5] Sovereignty is not just a matter of having
effective control over someone or something, nor is it just the same as
having absolute power over someone/something. It is actually more
concerned with deciding what is to be recognized as either someone

or just defined as something (of no political consequence). And this is Agamben's (1998, 181) point: "The fundamental activity of sovereign power is the production of bare life as originary political element and as a threshold of articulation between nature and culture, *zoē* and *bios.*

Some aspects of Eckersley's argument might also be presented in what at first appears to be a similar (if inverted) light: not in terms of control or authority per se but in suggesting that more things *should* be officially recognized (by the nation-state) as matters for ecopolitical decisions rather than as belonging to a globally depoliticized market. Here sovereignty is supposed to operate as the principle invoked to justify and the actual decisions made to progressively alter the dynamics of the balance between what counts as political and what counts as an apolitical economic exchange. But again, the point is that it is not the power of sovereignty that makes things (for example, certain kinds of economic transactions) *political* (at least not in Arendt's sense); quite the contrary, this is what politics as such does. Sovereignty is, at best, parasitic upon politics as such and only invoked to monotheistically suspend further political debate about whether certain economic matters are or are not political and especially to justify those decisions on which the nation's very survival is supposed to depend—including the decision to use force, when "necessary," to make sure all recognize and comply with the nation's (as the privileged political unit) interests.[6] In other words, the only thing that can result from reliance on a principle of sovereignty (rather than the encouragement of ecological politics as such) is not the repoliticization of an economically led biopolitics but its replacement (on the grounds of survival) with an ecological biopolitics.

And, despite Eckersley's erstwhile sensitivity to the complexities of political discourse and the unusual attention she pays to the possibilities of giving political voice to ethical concerns for the more-than-human world, this danger remains wherever the prerogatives of sovereignty are considered the ultimate political principle. Where the radical ecologist (and the savage democrat) want to argue that nothing can, or should, *as a matter of principle,* be drawn outside of ethico-political bounds, and the free-market enthusiast thinks that every aspect of the entire world (except, perhaps, those defined as properly human) is reducible to an economic value dictated by a market artificially kept free of politics, Eckersley wants to institute a compromise. But this compromise takes the form that "ecological freedom *for all* can only be realized *under* a form of government that enables and *enforces* ecological responsibility" (Eckersley 2004, 107, my emphasis). There is, perhaps, more than an echo of Hobbes (or Orwell) here. But more important, such statements misunderstand both how (ethical and political) responsibilities arise and what "ecological freedom for all" requires, which is precisely being free from the threat of the a priori reduction of populations to bare life and/or standing reserve, that is, the threat of sovereign (and anthropological) decisions.

Eckersley's compromise leaves open the fact that this demarcation line can—and *will*—be drawn somewhere by those states claiming the prerogative to do so and the pragmatic consequence that as these decisions are translated via individual nation-states into an international or supranational framework (an architecture of "global environmental governance" [5]), they become increasingly distant from, and less responsive to, the political communities and individual actions that are the only creative sources of any ecological politics.

To base an argument for a greener world on the deployment of a nation's sovereign powers, rather than in ecological politics per se, is to make it subservient to a state's constant ability to reduce political life to bare life and ecological life to standing reserve. Green sovereignty would be no less dependent on the state of exception nor on its enactment in general form as the biopolitical management of human (and nonhuman) populations, something all too easily associated with the cybernetic premises of weak or strong forms of ecological modernization and the technocratic exceptionalism espoused by Lovelock. Green states could also take many forms that are far from conducive to facilitating a political ecology—and state sovereignty gives no basis for criticizing these forms, since noninterference in internal state matters was, after all, one of the original bases of the Westphalian notion of sovereignty.

Ultimately, *the argument for green sovereignty is an argument against green politics.* This certainly is not the intention of those espousing the merits of a green state, and it does not follow that state involvement in ecological concerns is always destructive. However, it does mean that arguments for a radical ecological politics have to be clearly separated from, and opposed to, arguments for green sovereignty— something equally true where this same model of sovereignty provides a model for transnational governmental bodies. There may sometimes be short-term tactical advantages in having states oppose the depredations of global capital or, for that matter, supporting transnational tendencies that might make states less ecologically destructive (as, for example, the political transformation of the International Whaling Commission has done for a short time). But this achievement is always dependent on political pressures, and there is no general model or process that can replace this need for ethicopolitical involvement. If humanity is to avoid being reduced to bare life, and the living planet is not to become a bare rock, then we should not confuse politics with process or the creative constitutive political power of Earth's inhabitants with the constitutional power of the state to suspend such politics.

[handwritten: → inhabitant]

The Denizen: Being-with-Others, Worldly Responsibility, and Ecological Community

[handwritten margin note: we share the world with other denizens whom we ha little in common with (animals, plant sect)]

[handwritten margin note: morality is our common link w/ other denizens]

The unity of a world is not one: it is made of a diversity, including disparity and opposition. It is made of it, which is to say that it is not added to it and does not reduce it. The unity of the world is nothing other than its diversity, and its diversity is, in turn, a diversity of worlds . . . its unity is the sharing out *[partage]* and the mutual exposure in this world of all its worlds. The sharing out of the world is the law of the world. The world does not have any other law, it is not submitted to any authority, it does not have any sovereign. . . . The world is not given. It is itself the gift. The world is its own creation (this is what "creation" means).

—Jean-Luc Nancy, *The Creation of the World or Globalization*

Global capitalism represents a situation in which everything—the whole world—is reduced to economic values, to monetary equivalence. But where everything is so reduced, what value does the *real* world have? It also represents a situation in which everything is increasingly biopolitical and where, within these forms of governance, everyone is presupposed to be equally guilty of participating in the environmental holocaust now unleashed. But again, to paraphrase Arendt's remark, where everyone is guilty—and, it might be added, where everything is (bio)political—no one is guilty and nothing is political. And so we—those who concern themselves with (changing) this situation—are seemingly left with only a questionable reality (the real world), with ethical responsibilities that belong to no one in particular, and with nothing that can be defined as political. And if we reject the ever-present temptation to posit alternative first principles *(archē)* as a more fundamental ground of critique, then this is all we have with which to create provisional (nonfoundational and sustainable) polities, to re-conceptualize ethical, political, and ecological, communities.

But, perhaps this is just what we need. And putting the situation so starkly actually reveals something of the dangers and the possibilities of this global environmental crisis (this state of emergency that is already becoming normalized). For example, the questionability of the real world (What is real? What value does it have?) reflects back on the realities of the global economic system. It becomes obvious that, even in its own restricted terms, the values constituted by the economic realities of globalization cannot possibly be all that matters. (Indeed, they no more capture the real world than the concept "dog" can bark, although this is certainly not to say that the circulation of money and concepts do not *affect* [kinds of] reality.) Imagine presenting the world's inhabitants with $200 quadrillion for the entire planet—in just what sense is the real problem here the *inadequacy* of money offered? If one made the offer $300 quadrillion, would that put matters right? Would it then become a good deal?

It would be ethically, politically, and ecologically astute to reject any such deal, and once put in such terms, the stupidity of the suggestion becomes obvious. Yet the biopolitical reduction of the world, the strip-ping of its ethical, political, and ecological possibilities is driven by this same "restricted economy" (by an economy of monetary equivalence, of the Same), and by this failure to recognize that there is always a gap between representation and reality, there is always much more to what matters than money. From a radical ecological perspective, *the world is not for sale.*

And this, too, is why ethical and political responsibilities emerge through a realization of the ultimately unrepresentable and unex-changeable singularity, the infinity of Other beings (their continual questionability)—a recognition of Earth and its inhabitants' exces-sive nature *(physis)* and their wildness. And this infinity is tempered only by the finitude of each being's existence, by its mortality, the fact that their world ends, that my world will end, that the whole world will end, sooner or later. Our sharing in this world is based, then, on having nothing in common in both senses of this phrase: there are no essential commonalities that can define an ethical/political/ecological community, only differences and diversity, only our singularity and our being together in the face of nothing, of death (which is not a part of life), of ceasing to exist, an end that comes to us all. And as the hunters at Lascaux apparently knew, responsibility and ethics emerge in recognizing the impossibility of an exchange that can never actually compensate for taking another's life in this world, however necessary it may be for survival and however inescapable a part of everyday life it may constitute. This knowledge puts an end to innocence.

This notion of a provisional politics might be better understood through a comparison with Jean-Luc Nancy's (2007) similarly nonfoundational and self-sustaining approach. Nancy adapts Marx's critique of world capitalism in order to illustrate a distinction between economic globalization—the current worldwide "inter-dependence of the exchange of value in its merchandise-form (which is the form of general equivalency, money)" (Nancy 2007, 37)—and what Nancy terms *mondialization*—humanity's "world-forming" activities. For Marx, global capital extorts the value of human labor but nonetheless inevitably reveals, despite (or rather through a critique of) its fetishization of the commodity form, the "real value of common production" (37), that is to say, the ways in which humans actually gather together to historically *produce* a world of value. Global capitalism thus comes to be seen in the light of world history, and only following this recognition of the real source of value (a recognition Marx's own work claims to bring to fruition) "will the separate individuals be liberated from the various national and local barriers, be brought into practical connection with the material and intellectual production of the whole world and be put in a position to acquire the capacity to enjoy this all-sided production of the whole earth (the creation of man)" (Marx, *The German Ideology,* cited in Nancy 2007, 36).

Nancy argues that this emphasis on humanity's world-forming activities begins to offer a broader perspective from which to critique the world-destroying effects of globalization—its creation of what he terms an un-world *(immonde)*—and, as he also points out, it has the advantage of focusing attention on "this" world and not another world or a theologically derived "beyond-the-world" *[outre-monde]* as the only possible source of values: "The place of meaning, of value, and of truth is the world" (2007, 37). Values are immanent to the world's formation, not something added or guaranteed by some external ordering principle *(archē)*. However, Nancy argues, this same passage can also be read as presenting a post-Hegelian humanist eschatology in which the "human being, as source and accomplishment of value in itself, comes at the end of history when it produces itself" (38). That is, on this reading, Marx regards communism as that end in which the "real values of our common production" (37) are revealed as such and liberated from extortion by and through the production of a common humanity.

Like Bataille, Nancy rejects this quasi-theological eschatology because (as with Hegel's dialectic) it too suggests "an accomplishment without remainder" (38) and as such typifies a restricted economy and an understanding of (the end of) world history as an all-encompassing totality. This envisaged end, while perhaps more realizable than its otherworldly counterparts, is so only to the extent that it engages in what Lefort too (see chapter 6) would regard as the totalitarian fantasy of eventually overcoming all those differences that matter (ethically and politically). Despite Marx's resolute refusal to engage in utopian speculations, this totalizing fantasy of a community without ethical and political differences may, Nancy suggests, be precisely what leads *this* understanding of communism to take totalitarian forms. "But what, since Marx, has nonetheless remained unresolved *[en souffrance]* . . . [is] the world of proper freedom and singularity of each and of all without claim to a world beyond-the-world" (38).

Leaving aside questions regarding Marx's actual understanding of the end of history (which are primarily of interest to those claiming to be his true heirs and upholders of his infallibility), Nancy's argument suggests a very different interpretation of the relation between world forming and the world's end. Rather than thinking of the end of the world teleologically, as eschatology or historicism do, as the completion of and the final word on world history, he thinks of it in terms of the recognition of the world's finitude, its (and our) eventually ceasing to exist.

We might say that the idea of a world without end makes sense only in terms of this possibility, that is, choosing to abandon teleology. A provisional politics can have no ultimate end, no conclusion; and no moment of revelation will ever make sense of world history as a whole. Meaning and sense, too, are provisional and created through this sharing of a world, a provisionality that erodes any possibility of passing final judgment. So when Nancy writes explicitly of "the end of the world" (1997, 4), he means the recognition that there is no endpoint, no completion of history, philosophy, politics, and so on. He is speaking of *"the end of the sense of the world"* (5), that is, the world understood as a cosmos the sense of which is pregiven from the beginning and/or revealed at the end. This makes sense, too, from an ecological point of view so long as Nancy's anthropocentric understanding is also

supplemented with the recognition of the end (the finitude and extinction) of many of the other(s') worlds that together constitute the world.

A provisional ecopolitics chooses to recognize that we live in a world
the end (mortality) of which is certain, even if that end may be post-
poned until the expanding sun one day dissolves this planet. Of course,
the initial recognition of our own mortality, a forewarning that our
(phenomenal) world, our being-here, will one day cease to exist, and by
extension the finitude of the whole world, is also a precondition of (and
a motivation for) the positing of millennial and otherworldly eschatolo-
gies. For example, dominion theology's rapture (see chapter 1) is just
such an escapist fantasy, a fearful reaction to the recognition of mortality and the inevitable transience of all sense, meaning, and values
created in and through this world. Even the rock paintings at Lascaux
already reveal both that initial awareness of being suspended over the
abyss of death and some form of belief in another world, that of the
dead, to which only the shaman can travel and return. And, although
it is impossible to overestimate the importance and the culturally mediated variety of responses to (and denials of) such a recognition,[7] Nancy
(1997, 41) asks whether the explicit recognition that despite all these
past variations, actually "nothing else was at stake, in the end, than
the world itself, in itself and for itself," might itself provide sufficient
material for the creation of political communities.

As has already been argued, any ecological politics is inevitably
and paradigmatically a this-worldly response to just such a recognition (after all, the world of the Auroch is extinct; all but its image has
gone forever). But this raises further questions concerning the nature of
Nancy's world-forming communities, most obviously, one he asks himself: "What is a world? Or what does the 'world' mean?" (1997, 41), but
also "Who/What forms it?" Nancy's (2008) answer, developing a post-
Heideggerian phenomenological approach, is that the world is a form
of "being-there-with" *(Mit-da-sein)* others, where each of these terms
(being, there, with, others) matters and informs the others. "Being-
with" should not be thought in terms of having some*thing* in common
(of any essentialism), still less of making this "thing" a principle *(archē)*
of "ethicopolitical" community (as the anthropological machine does
and as the totalitarian danger of Heidegger's own Nazism illustrates).
And it should not be thought in terms of a relation among individuals that assumes each being's preexistence as an already individuated
being (as liberal individualism, for example, does).

Whether or not, as Marchant (2007, 79) argues, "Mitsein ontologically precedes [or is coemergent with] Dasein," the *"between,* the *with,* and the *together* are all irreducible aspects of being—which therefore can only be thought of as being singular plural." This "being singular plural" (Nancy 2000) has many similarities to Agamben's (2001) understanding of "whatever" being, of "being such as it is" (see chapter 2) insofar as it too is potentially constitutive of an emergent political community.[8] From Nancy's perspective, a single being would be a contradiction in terms. And, in this sense at least, Nancy echoes Levinas concerning the precedence and inescapability of relations with others, although for Levinas this relation is preontological, necessarily ethical, and relates to a solitary Other rather than the broader multitudinous resonances of sharing a world (being-with-others) that Nancy invokes (see Watkin 2007).

The world is a sharing in creation, "not thought in substantivist terms of something that is owned and shared by already constituted owners and sharers, but as an act of finite being" (Caygill 1997, 22). The world is not a resource to be shared (divided out) among us; the world is the appearance of the "between us" created through our interactions; our involvements; our resonances and dissonances; our collisions, conflicts, and concerns; our mutual exposure—our coappearance in/as the world. The singularity and unrepeatability of a life is this excessive being-(outwith)-others. And the world is created, as is any community, through the constitutive acts of those singularities who, together, compose it, ex nihilo, from and in the face of "nothing."[9]

> A world: one finds oneself in it *[s'y trouve]* and one is familiar with it *[s'y retrouve];* one can be in it with "everyone" *[tout le monde].* . . . A world is precisely that in which there is room for everyone: but a genuine place, one in which things can genuinely take place (in this world). (Nancy 2007, 41–42)

The world is an unintended and unexpected gift that offers singular possibilities of presently finding oneself with others. But this means that it is not "given" in the sense of being the unalterable foundation of our being, although we cannot grow, apart from our being involved in it. We coevolve as parts of the world, as a diversity of worlds in correspondence (Nancy says we "com-pear" *[com-paraît]*) with each other, and so far as is open to us as ethical beings, this correspondence and coappearance can be imbued with a sense of responsibility for Others. To take responsibility requires inhabiting the world in terms of occupying a self-standing place in relation to others ("standing," as Nancy points out, is at the root of the words *ethos* and *habitus*—see also Smith 2001a). "A world is [or at least can be] an ethos, a *habitus* and an inhabiting" (Nancy 2007, 42). This self-standing is both where we come from and what is created as we act or speak, when our singularity before others is expressed, thus constituting a (political) world with others. This creation is always already in correspondence (which again, as Murdoch, Levinas, and Arendt argue, does not mean agreement) with others. Singularity is this exposure of the existent to others in the world. Our political singularity is revealed as we act; we are exposed as *who* we are, before others. *How we act reveals who we are*

"World-forming," as the coappearance and cocreation of singular plural beings, provides a very different understanding from that which regards the world as an unalterable overarching externality, and especially from that subjection of oneself to the sovereignty of the world as an inevitable process, fate, Nature, human nature, teleology, tyranny, historicism, and so on, since this is to *give up* one's (ethical and political) responsibilities and one's singularity—to become bare life, perhaps even a (human) resource. It also provides a very different understanding from that archetypal mistake of thinking the world as something given to us (on the basis of some unchanging first principle, whether God, natural rights, the special value of human labor, and so on) for us to rule, steward, manage, or make subject to our sovereign powers; and this, of course, in its (bio)political form, is the route taken by economic globalization.

But this again raises the questions of (world) sovereignty and of the anthropological machine. Nancy is intensely critical of the principle of sovereignty at the level of the political community because it denotes that power to decide the end of politics—in both the sense of defining the ends to which politics is to be put (for example, by the sovereign declaration of a state of war) and of putting an end to politics as such: "That which is sovereign is final, that which is final is sovereign" (2000, 120).[10] Sovereignty, in these multiple senses *denotes the end(s) of the political world,* a world that can only really be sustained as (in Agamben's terms) "pure means." "The problem is not a matter of fixing up *[aménhanger]* sovereignty: in essence, sovereignty is untreatable" (Nancy 2000, 131).[11] Nancy also recognizes, in a way that parallels Agamben, that where economic globalization is concerned, sovereignty is not usually explicitly invoked. The political is instead replaced by a biopolitical system in which decisions are deemed to be matters of technical (apolitical) interventions and "progressivity takes for itself the controlled management of natural life" (Nancy 2007, 94). This is what Nancy calls "ecotechnology." The problem with globalization is that (drawing on Heidegger's analysis of the technological enframing of the world—see earlier discussion) "what forms a world today is exactly the conjunction of an unlimited process of an eco-technological enframing and of a vanishing of the possibilities of forms of life and/or of common ground" (95).

There are many ways to articulate possible links between Nancy's radical materialism and a provisional ecopolitics precisely because Nancy's emphasis, unlike Marx and the early Heidegger, is not always on the world as the creation of human beings. He does not share Marx's productivism (see Smith 2001a), arguing explicitly for what he terms the "transimmanence" of the world, a creativity not reducible to any economic (instrumental) productivism nor to a world represented as awaiting humanity's transformative powers. "There is not, on the one side, an originary singularity and, on the other, a simple being-there of things, more or less given for our use" (Nancy 2000, 17–18). And while Nancy still underestimates the extent to which this world and this creativity, this being-there-with-others, is constituted by and with more-than-just-human beings, he also recognizes the ways in which "the others who we are always already with are many in number, and not all human in kind" (Bingham 2006, 492). Such "thinking," argues Nancy (2000, 17), "is in no way anthropocentric; it does not put humanity at the centre of 'creation'; on the contrary, it transgresses *[traverses]* humanity in the excess of the appearing that appears on the scale of the totality of being, but which also appears as that excess *[demesure]* which is impossible to totalize." In other words, even though it is human being-there that offers the possibility of recognizing the finitude of the world as a whole, we must avoid "giving the impression that the world, despite everything, remains essentially 'the world of humans'" (18). Rather, *"humanity is the exposing of the world; it is neither the end nor the ground of the world; the world is the exposure of humanity; it is neither the environment nor the representation of humanity"* (18). And so, while Nancy does not offer either an environmental ethics or an ecopolitics, his thinking allows, and perhaps even encourages, such possibilities.

It might be fair to say that Nancy's world is anthropocentric in the same sense that Evernden's radical ecological phenomenology (see chapter 5) is necessarily anthropocentric, that is insofar as the world emerges as the reality of a human experience of being-in-the-world. But the more-than-human world existed before, and still exists out-with, any human existence, although it might or might not have been apprehended as "the world." This nonteleological creativity of the be-ings that share and compose a more-than-human world is called evolu-tion. As Darwin recognized, humanity is not the end of an evolutionary process, not its telos nor its end product. (Nature, after all, is not a factory.) And the difference and diversity of the worlds that compose the world is beyond the wildest imaginings of any human being (think of the world of the tick—see chapter 4). This diverse sharing is what evolution is—a matter of differing involvements, apportionings and appropriations—but it is not a sharing according to any principles of justice.[12] Evolution is not subject to the sovereignty of the Good or of politics. Insofar as the world is *without* (lacking) ethical and political beings, it remains entirely innocent and apolitical. But insofar as it is *outwith* ethical and political beings, it will always potentially be a mat-ter of ethical and political contestation. And if, as Darwin supposed, a grain of sand can turn the evolutionary balance (which is in no sense the same as the balance of justice), then today each ethical and po-litical act can bring ecological communities to the point of collapse or, alternatively, constitute a place for them to sustain themselves.[13] Even the most insignificant ecopolitical act might turn the balance between world formation and the unworld of globalization.

And this situation, in which globalization threatens to destroy en masse the world's diversity, to institute and institutionalize a world-wide ecological holocaust, is where we stand today, where our ecologi-cal responsibilities come from. We have to re-create an ecological com-munity able to withstand the biopolitical dangers in acquiescing to a principle of human sovereignty (dependent on the anthropological ma-chine) that declares that the more-than-human world, the world out-with humanity, is entirely subject to, and yet simultaneously excluded from, the realm of ethics and politics. We need to resist the reduction of the natural world to a resource to be governed. Meanwhile, we also need to stand up and speak out against the normalizing of a state of (ecological) emergency that reduces our ethical and political being to bare life. The possibility of saving the political world and many (perhaps most) more-than-human worlds might well depend on our abilities to create diverse forms of ethically informed provisional politics capable of giving some shelter and sustenance to others by engaging in an anarchic, endless critique of sovereignty in all its guises.

These ecological communities, these sharings of the world that are evolutionary, ethical, and political, human and more-than-human, cannot be envisaged as an end *(telos)* without falling back into teleology or myth. An ecological community is no paradise but rather the coappearance of living ethical/political/ecological realities—that is this-worldly realities—appearances that, as Arendt argues, are what constitute the possibilities of freedom and also what preclude any Kantian pipedream of a worldwide state of "perpetual peace" (see Caygill 1997, 27). The opposite of perpetual peace is not necessarily perpetual war, but a shared being-with-others that each ethicopolitical being can, at least to some extent, choose to make more, or less, peaceful for others, even more-than-human others.

This is how an ecological community can be understood. And where one ecological community begins and ends is just as impossible to define as the beginning and end of humanity. Its constituents cross and recross any boundaries that might be imposed. The dragonfly, for example, is a denizen of the pond. First (or is it last?) as an egg, laid with others in shallow water (perhaps a rainwater puddle) and covered with protective sticky jelly. Once hatched, it inhabits the pond's watery medium as a nymph, a larval predator of almost anything it encounters. It might remain there for up to eight years and only then, if it has not fallen prey to another of the pond's denizens, begins another almost complete transformation. It starts to breathe air and then broaches the pond's surface and climbs, eventually attaching itself, perhaps to the side of a tree, where its metamorphosis takes place. It sheds its old skin, leaving behind its last exuviae. As dawn breaks, it takes flight for the first time, now a denizen of the air, capable of spectacular acrobatics. It lands on an outstretched hand. It will live at most for a few months, but even so, some dragonflies can migrate to very different ecological communities where they are no less at home. Some even cross the Gulf of Mexico. Are human beings less adaptable?

And if the stone is, as Heidegger claims, worldless (in the sense of lacking its own phenomenal opening into the shared world), it no less constitutes a part of the worlds of others, a part of the ecology of the world. It may still be a thing of singular concern. It matters. Every grain of sand can matter, and in many different ways. The more-than-human world does not have any sovereign power to decide what should and should not matter. Of course, these singular concerns too are diverse, more or less involved, and only rarely, if at all, ethical. The snake covets the rough-grained warmth of the granite rock. The turtle draws itself out onto the ledge. The gray heron stands perched, waiting. Their concerns are indeed beyond our ability to fully grasp or represent, but that is true of all others. They still share in the creation of the world outwith us and are not altogether beyond the touch of our ethicopolitical concerns.

There may be ecological struggle, but also cooperation; competition, but also symbiosis; fear, but also enjoyment; life, but also death. In a single walk along a "tangled bank" (Darwin 1884, 429) or through a woodland, we encounter all of these events. The only event in our neighborhood that necessarily eludes us is our own inevitable death. And so long as we live, those capable of so responding can regard the creations of evolution, the "endless forms most beautiful and most wonderful" (429); the fellow constituents of a world who can, when we least expect it, take us out of our self-centered obsessions, and allow us to experience and to consider what it might mean to share in the creation of such a world rather than be the domineering agent of its destruction. *We can all be denizens of this world.*

APOLOGUE

In Relation to the Lack of Environmental Policy

THE PURPOSE OF THIS BOOK is to open possibilities for rethinking and constituting ecological ethics and politics—so should one need to apologize for a lack of specific environmental policies? Should the book declaim on the necessity of using low-energy light bulbs or of increasing the price of gasoline? Alternatively, should one point out that such limited measures are an apology (a poor substitute) for the absence of any real ecological ethics and politics? Is it possible to read this book and still think that I believe the problem is one of the incompleteness of the current policy agenda and that the solution to our environmental ills is an ever-more complex and complete legislative program to represent and regulate the world?

It may be possible, but I hope not. For this desire for legislative completeness, this institutional lack (in a Lacanian sense), the desire for policy after policy, is clearly the regulative counterpart to the global metastasis of those free-market approaches that effectively reduce the world's diversity to a common currency, a universal, abstract, monetary exchange value. They are, almost literally, two sides of the same coin, the currency of modernism and capitalism, and their presence is a tangible indicator of the spread of biopolitics. Meanwhile, the restricted economy of debates around policy priorities and cost–benefit analyses almost always excludes more profound questions and concerns about the singular denizens of a more-than-human world. The purpose of this book is to provide philosophical grounds on which such questions and concerns can be raised, to challenge the myths and metaphysics that would regard them as inconsequential. In this, no doubt, it is already overambitious.

In any case, unlike the majority of people writing on the environment, I do not have a recipe for saving the natural world, a set of rules to follow, a list of guiding principles, or a favorite ideology or institutional form to promote as a solution. For before all this, we need to ask what "saving the natural world" might mean. And this requires, as I have argued, sustaining ethics, and politics, and ecology over and against sovereign power—the exercise of which reduces people to bare life and the more-than-human world to standing reserve. This is not a politically or ethically neutral position in the way that liberalism, for example, would like to present itself; it reenvisages ecological communities in very different ways.

Of course, I have opinions on what is to be done, although not in any Leninist fashion. My sympathies might find some expression in, for example, ecologically revisioning Kropotkin's mutual aid and Proudhon's mutualism. But expanding on these ideas here would just provide an excuse for many not to take the broader argument about sovereignty seriously. What we need are plural ways to imagine a world without sovereign power, without human dominion. And so, instead of an apology or an apologia for the lack of policy recommendations (and who exactly would implement them), I offer an apologue, a "moral fable, *esp.* one having animals or inanimate things as its characters" *(New Shorter Oxford Dictionary)*.[1]

I have in mind a recent image that momentarily broke through the self-referential shells that accrete around so many of us, cutting us off from the world as it really is. Not an ancient painting on a rock wall, but a still photograph of a living polar bear standing, apparently stranded, on a small iceberg—a remainder of the ice floes melting under the onslaught of global warming. Now only a wishful thinker (or a bureaucrat) would contend that such bears will be saved by more stringent hunting permits (deeply as I abhor sport hunting), by a policy to increase ecotourism, or by a captive breeding program in a zoo. These measures are clearly inadequate for the task. Despite conforming to a policy model and taking account of realpolitik, they are far from being realistic. For the bear, in its essence, is ecologically inseparable from the ice-clad north; it lives and breathes as an inhabitant, a denizen, of such apparently inhospitable places. It is an opening on an ursine world that we can only actually imagine, but its image still flashes before us relating "what-has-been to the now" (Benjamin 1999, 462).

This image is one of many that, through affecting us, have the potential to inspire new ethical and political constellations like those of radical ecology. For a radical ecologist, the bear is not a resource (not even an ecotourist sight) but a being of ethical concern, albeit it in so many respects alien, pointless, independent from us—and, for the most part, indifferent to us. It can become close to our hearts (which is not to expect to hug it but to say that it elicits an ethical response that inspires a politics). And this politics argues that only a hypocrite could claim that the solution to the plight of such polar bears lies in the resolution of questions of arctic sovereignty, in an agreement to take account of the rightful territorial claims of states over these portions of the earth.

Once defined as sovereign territories, there can be no doubt that the minerals and oil beneath the arctic ocean will be exploited in the "name of the people and the state" and to make money for capitalists. And this will add immeasurably to the same atmospheric carbon dioxide that already causes the melting ice. After all, there is no other purpose in declaring sovereignty except to be able to make such decisions. And once this power of decision is ceded, all the nature reserves in the world will not save the bears or the ecology they inhabit. They will (in Nancy's and Agamben's terms) have been "abandoned," "sacrificed," for nothing and by nothing.

So what *could* be done? When those seeking a policy solution to the Arctic's melting ice ask this question, they are not looking to institute policies that would abandon or radically transform capitalism, abolish big oil, or even close down the Athabasca tar sands. They expect to legislate for a mode of gradual amelioration that in no way threatens the current economic and political forms they (wrongly) assume to be ultimate realities. In short, they want a solution that maintains the claims of ecological sovereignty.

But perhaps we could, at least as a point of discussion, suggest a model in which such sovereignty is suspended, such as that found at the other end of the earth, in Antarctica. Of course, Antarctica does not provide a perfect solution, but it does offer an example, however inadequate, of the institutionalized "suspension of ecological sovereignty" that could easily be extended if any governments or policymakers had the inclination. We might, as evidence for this assertion, naïvely take these states' own words for this. For example, the British Antarctic Survey states:

There are few places on Earth where there has never been war, where the environment is fully protected, and where scientific research has priority. The whole of the Antarctic continent is like this. A land which the Antarctic Treaty parties call a natural reserve, devoted to peace and science. The Antarctic Treaty came into force on 23 June 1961 after ratification by the twelve countries then active in Antarctic science. The Treaty covers the area south of 60°S latitude. Its objectives are simple yet unique in international relations. They are:

- to demilitarize Antarctica, to establish it as a zone free of nuclear tests and the disposal of radioactive waste, and to ensure that it is used for peaceful purposes only;
- to promote international scientific cooperation in Antarctica;
- to set aside disputes over territorial sovereignty.

The treaty remains in force indefinitely.

How fabulous! And of course, this description is fabulous, for Antarctica is no modern state of nature, nor is it in an entirely natural state. But then, as this book argues, no state of nature has ever actually existed. So it should not surprise us that the language of use and of the nature reserve is still in play here or that human dominion is still presumed without any justification other than that of attaining scientific knowledge from a wilderness which, although treeless, is still made to sound almost paradisiacal. And, of course, we know that militarization and resource exploitation are never far away, that pollution still occurs, that global warming's effects are strongly felt here, that there are plenty of policies on hunting, and that biopolitical trappings will no doubt continue to proliferate under this "treaty regime" as each nation jostles for control of commercially exploitable resources.

And yet here is an imperfect real-world example of the "suspension of [at least national] sovereignty";[2] of the (purportedly) indefinite suspension of the power to make a decision that turns entire continents into standing reserve. To the extent that this is made apparent, its ecological consequences are almost entirely benign for all concerned. Antarctica's more-than-human inhabitants are, to this degree, released into their essence, they are let be in the sense already discussed (see chapter 4). Where they are not, then radical ecologists (whether in the form of Greenpeace or the courageous and somewhat ironically named Sea *Shepherd*) act to make this an ethical and political matter. And this includes applying pressure to those trying to use the global sovereignty of objective scientific knowledge (as Japan's whalers do) as an excuse.

Unfortunately (but hardly unexpectedly, given the oil corporations' interests in Alaska), the United States rejected the idea of an Arctic Treaty put forward by the World Wildlife Fund (2008) and the negotiations to this end proposed by the European Parliament, claiming that a treaty "along the lines of the Antarctic Treaty—is not appropriate or necessary" (Anderson 2009, 105). This leaves only the Law of the Sea in place, and this specifically allows (under article 76) for the making of claims to sovereign territory over an "extended continental shelf," hence the bizarre symbolism of the Russian Federation planting a rust-free titanium flag on the sea floor at a depth of 13,980 feet. As for the seabed beyond these claims, it is currently considered to be "the common heritage of mankind" (Howard 2009, 18)!

Perhaps the power of the Antarctic Treaty, if it has any, lies in its potential to be turned toward a weakening of the metaphysical ideal of sovereignty and dominion altogether. For to expect that nation-states will simply extend and radicalize the example of the Antarctic, first to the Arctic and then to the wider world, would be rather like expecting the Marxist state to automatically wither away. In any event, radical ecology puts no store in automatic processes, still less in any state's willingness to give up its sovereignty. Its hope lies in the creativity of ethical and political action and in the world understood as that ecological creation in which everyone shares.

NOTES

1. Awakening

1. It is now postulated that the apse may actually have been the site of a second entrance to the cave rather than one of its more remote recesses. Because of differences in its style of composition and in the pigments used, the adjacent figure of the rhinoceros is thought to be unrelated to the other elements of the painting (Anjoulet 2005).

2. This shamanic interpretation has considerable support, as exemplified in David Lewis-Williams's (2002, 264–66) discussion in *The Mind in the Cave*. Lewis-Williams, though, appears unaware of Bataille's much earlier writings on Lascaux. If the figure does represent a person in shamanic trance (which the erect phallus might also support), it complicates but still supports Bataille's interpretation. In any event, the relationship depicted should not be understood simplistically as the quid pro quo of a human death (actual or ritual) for an animal death; it is not a trade but, in Bataille's sense, a "religious" and hence mysterious, metonymic affinity, an equivalence in the face of death.

3. "Animals wait for nothing, and death does not surprise them; death in some way eludes the animal . . . man's intellectual activity put him in the presence of death, in the radical terrifying negation of what he essentially is" (Bataille 2005, 152). "The mortals are human beings. They are called mortals because they can die. To die means to be capable of death as death. Only man dies. The animal perishes. It has death neither ahead of itself or behind it. Death is the shrine of Nothing, that is, of that which in every respect is never something that merely exists, but which nevertheless presences, even as the mystery of Being itself" (Heidegger 2001, 176).

4. Gianni Vattimo (2004, 27–28) suggests that Heidegger's term *Verwindung* indicates the possibility of weakening the hold of metaphysical absolutes (such as a concept of human nature) over our thinking, an approach he refers to as "weak thought"—*Il pensiero debole*. This "twisting" is necessary precisely because we cannot entirely overcome the legacy of metaphysics, since its effective historical presence haunts every mode of Western thought and never more so than today.

5. For Benjamin, historicism naturalizes the past as universal history, an inevitable and intractable process of progress or decline, leading up to and beyond the present moment, rather than as the constantly reworked and remembered material of human actions. For historicism, the paintings of Lascaux would stand as an "'eternal' image of the past" rather than as offering the possibility of "a unique experience with the past" (Benjamin 1973, 254). As Kendall (2005, 14) remarks, Bataille claims, albeit in a slightly different sense, that "prehistory is universal history par excellence."

6. Bataille studied under Alexandre Kojéve, the foremost interpreter of Hegel's philosophy and, as Derrida (1995a) shows, Hegel's influence on Bataille is pervasive.

7. Bataille (2005, 54) frequently suggests that for "men of primitive times, as for men of the modern day who we rightly or wrongly call primitives," there are shared patterns of belief that separate "them" from "us"—assuming a complicity between author and reader. Yet he also claims that "modern primitives," unlike "real primitives," lack "the upsurge of creative awakening that makes Lascaux man *our* counterpart and not that of the aborigine" (159, my emphasis). Either way, Bataille relegates contemporary aboriginal peoples to an indeterminate status in a way that perfectly illustrates what Agamben regards as the political dangers of the ways in which the anthropological machine produces zones of indeterminacy where certain populations are treated as less than fully human.

8. Agamben (2004, 34) makes this same point in his discussion of Haeckel's positing a missing link between animal and human, which he called the ape-man *(Affenmensch)*.

9. "Where Hegel relayed the history of consciousness, Bataille reveled in all that consciousness cannot capture, that words cannot describe." (Kendall 2007, 208). If, for Hegel, "the wounds inflicted by history invariably heal without leaving scars" (Gourevitch 1988, ix), then Bataille frequently described his own life as an open wound.

10. It might also suggest that recent debates concerning posthumanism and hybridity stemming largely from Donna Haraway's (1991) work recapitulate aspects of Bataille's speculations about the condition of humanity at the end of history, in particular his invocation of the image of the headless, acephalic man (see later in this chapter). Of course, many of these debates focus at least as much on the issues raised by technology as they do on nature (Hailes 1999; Lippit 2000).

11. Adams also quotes lines from Pope's "Essay on Man": Man walk'd with beast, joint tenant of the shade; / the same his table, the same his bed; / no murder cloath'd him, no murder fed.

12. Derrida provides an extensive commentary on the ambiguous and mutually imbricated relations underlying the attempt to demarcate the human from "the animal" that begins with a discussion of the feeling of shame when standing naked in front of "his" cat who, of course, is also naked (unclothed) and yet not naked in the sense that Derrida (like Adam and Eve) *feels* naked. "The animal, therefore, is not naked because it is naked. It doesn't feel its own nudity. There is

no nudity 'in nature.' . . . Man could never be naked any more because he has the sense of nakedness, that is to say, of modesty or shame. The animal would be *in* non-nudity because it is nude, and the man *in* nudity to the extent that he is no longer nude. Thus we encounter a difference, a time or *contretemps* between two *nudities without nudity*. This contretemps has only just begun giving us trouble *[mal]* in the area of the knowledge of good and evil" (Derrida 2008, 5).

13. *Spare Rib* was the title of an influential feminist periodical in the 1970s and 1980s.

14. Of course, White was right that this biblical narrative does indeed haunt all subsequent Western understandings. We might note that even Feuerbach's rationalistic and radically demystifying reversal of this foundational myth, which recognizes how humanity creates its gods in its own image, challenges only the reality of God's earthly power, not the self-awarded status of human dominion that is, one might think, made all the more obviously arbitrary as this ideological veil is lifted. Ironically, White's paper would also be criticized by contemporary left-Hegelian successors of Feuerbach who, preferring a more "material" base to their history, sought to deny the causal powers of ideas but whose focus on relations of production and ideology critique still left the ultimately baseless structure of this claim to human dominion intact (see Hay 2002, 103).

15. See also Taylor (2007), although Taylor does not mention Vattimo.

16. Reading Passmore (1974, 29), one might, quite wrongly, be left with the impression that Black was straightforwardly a proponent of ecological stewardship. But he actually thinks it unlikely that even if an ideal of Christian stewardship could be defended as a valid textual interpretation of the Bible, it would provide sufficient motivation to bring about a change in contemporary ecological attitudes. This, Black (1970, 123) argues, might depend on extending ethical concerns to humanity in general, "dead, living or as yet unborn," an insightful remark given the role that "future generations" now play in ecological politics.

17. Gadamer (1980, 93–123) notes that there has been some controversy about the authenticity of this letter, but he is convinced that it is genuine and expresses Plato's views.

18. Although this rationalistic dialectic differs markedly from Benjamin's historical dialectics, there are interesting similarities in turn of phrase here. Benjamin describes the sudden flash when image and "the now" come together stating— "Only dialectical images are genuine images (that is, not archaic); and the place where one encounters them is language. Awakening." Benjamin was fascinated by Plato's theory of Ideas and their relation to language and history (Hanssen 1998, 24–25). He regarded "language as such" very broadly and nonanthropocentrically as the communicable expression of (material and hence historically transient) creation and not just as human communication—which is not to say that Benjamin is not anthropocentric when it comes to specifying the task of human language, which is naming things: "It is the linguistic being of man to name things" (Benjamin 1998, 110). For an extended discussion, see Smith 2001b.

19. Bataille also might have thought laughter an appropriate response because

for him, too, "laughter alone exceeds dialectics and the dialectician" (Derrida 1978, 256). Perhaps this is also why, as Arendt (1977, 82) remarks when discussing this same incident, Plato, "who argued in the *Laws* for the strict prohibition of any writing that would ridicule any of the citizens, feared the ridicule in all laughter." As Taminaux (1997, 1) notes, Arendt claims that Kant "seems to have been unique amongst the philosophers in being sovereign enough to join in the laughter of the common man." See Arendt (1977, 82–83).

20. Is it to defuse ridicule or to invoke laughter that Plato has the Stranger remark that the result of this process is "not without its interest for comedians" (1031 [266c]), for it apparently leaves humanity in the neighborhood of the portly, snuffling, pig. Gadamer (1980, 94), contra Arendt, claims that there is always an intentional "touch of humour and irony" in the way Plato differentiates concepts. "The *Parmenides,* especially, reads almost like a comedy."

21. Kalyvas makes an important connection between Foucault's account of pastoral power and Agamben's concept of biopolitics. However, he dismisses the political ramifications of the pastoral/stewardship metaphor for Plato, arguing that it comprises only the early stages of an argument leading to the introduction of a superior metaphor—that of *weaving* the political fabric of the city. Kalyvas thinks this weaving metaphor might ground a model of sovereignty quite different from that of the pastorate and of Agamben.

22. "Politics is the sphere of pure means, that is, of the absolute and complete gesturality of human beings" (Agamben 2000, 59).

2. The Sovereignty of Good

1. Despite its Greek roots, the word *ecology* was first coined by Ernst Haeckel in 1866 (Golley 1993, 2).

2. Murdoch here offers her own translation from Plato's *Republic* (592b).

3. The prime concern here is not whether Murdoch's interpretation of Plato's position is correct but how it operates to explain the intimate relation between ethics and seeing the world as it is. As Charles Taylor (1996, 96) states: "No one today can accept the Platonic metaphysic of the Ideas as the crucial explanation of the shape of the cosmos. And yet the image of the Good as the sun, in the light of which we can see things clearly and with a kind of dispassionate love, does crucial work for her [Murdoch]. It helps define the direction of attention and desire through which alone, she believes, we can become good."

4. Levinas makes another important distinction between "saying" *(Dire)* and "said" *(Dit).* If we attend to the saying (the Other's speaking to us) rather than just the explicit definable content of what is said, we hear that which escapes being fixed in language. From Levinas's perspective, philosophical analysis focuses on what is said; ethics, on the other hand, attends to that beyond the crystallization of existence in what is said. In a rather different way, then, for Levinas too, philosophy (or at least ethics as first philosophy) is also an attempt to say the unsayable.

5. This is also close to Heidegger's understanding of the role of philosophy

in relation to truth as appearance *(aletheia)* and nature *(physis)*—understood as a self-actualizing movement of both "revealing disclosure" pushing up into the phenomenal world and receding into the "sheltering enclosure" of the earth (Haar 1993). (Murdoch too might have noted this similarity had she completed the study of Heidegger she was writing toward the end of her life.) For Heidegger (1995, 187), philosophy circles the essence of its questioning; it is not a linear movement of progress towards a final definition. "The centre, that is the middle and ground, reveals itself as such only in and for the movement that circles it. The circular character of philosophy is directly bound up with its ambiguity, an ambiguity that is not to be eliminated, or, still less, leveled off by means of dialectic. It is characteristic that we repeatedly find in the history of philosophy such attempts to level off this circularity and ambiguity of philosophical thinking through the use of dialectic, and most recently in a grand and impressive form. Yet all dialectic philosophy is only the expression of an embarrassment."

6. Gadamer (1985, 131) claims "the mimetic presence that makes Plato's word and work so unforgettable emerges from this doubled and tense irony." He also states that one "can win a certain clarity by analyzing the argumentation of a Platonic dialogue with logical means, showing up incoherence, filling in jumps in logic, unmasking false conclusions, and so forth. But is this the way to read Plato, to make his questions one's own? Can one learn from him in this way, or does it simply confirm one's own superiority? What holds for Plato holds *mutatis mutandis* for all philosophy" (191).

7. As Wittgenstein (1990, 185 #6.4311) too remarks, "Death is not an event of life. Death is not lived through."

8. To think in terms of completion is to have forgotten or dismissed the inexplicable abyss that appears beneath our feet at the most unsuspected moments. The only response to this totalizing hubris is ironic laughter. To quote Derrida (1995a, 257) on Bataille: "The notion of *Aufhebung* (the speculative concept par excellence, says Hegel) is laughable in that it signifies the *busying* of a discourse losing its breath as it reappropriates all negativity to itself, as it works at the 'putting a stake' into an *investment,* as it *amortizes* absolute expenditure; and as it gives meaning to death, thereby simultaneously blinding itself to the baselessness of the nonmeaning from which the basis of meaning is drawn, and in which this basis of meaning is exhausted."

9. This elision, like the concept of sustainable development, is obviously a political compromise, since the Brundtland Report claims "ecology and economy are becoming ever more interwoven" (WCED 1987, 5) and that industrial development needs to change radically (213) because of its environmental consequences even as it supports a "new era of economic growth, one that must be based on policies that sustain and expand the environmental *resource* base" (1, my emphasis). The report sets out a model of human stewardship of natural resources for the benefit of current and future human generations, making only a passing allusion to environmental ethics (13).

10. Art is actually, says Murdoch (1970, 86), "less accessible but more edifying

than nature since it is actually a human product, and certain arts are actually 'about' human affairs in a direct sense."

11. Although Murdoch recognizes that historically this was an understandable response by modern philosophers faced with the collapse of a transcendent *super-natural* moral order.

12. This is an extrapolation from Murdoch's ethical position, not a statement about her own political views. While Murdoch's politics were originally strongly anticapitalist, she was a member of the Communist Party until at least 1942, and while part of her critique of existentialism was that, as a form of voluntarism, it was "the natural mode of being of the capitalist epoch" (1999, 224), she actually voted for the Conservative government led by Margaret Thatcher in the 1980s. Her biographer, Peter J. Conradi (2001, 499), suggests this was because she favored retaining an educational meritocracy, perhaps another echo of Plato's influence.

13. "The face disorients the intentionality that sights it. This is a challenge of consciousness, not a consciousness of the challenge. The Ego loses its sovereign coincidence with self, its identification where consciousness comes back triumphantly to itself to reside in itself" (Levinas 2003, 33).

14. Many biocentric theorists, including Mathews (2003), see systems theory as offering a way of grounding environmental ethics in an attenuated (cybernetic) model of selves as self-realizing systems and even of recuperating a similarly attenuated notion of telos or purpose in the world (for example, Morito 2002). For a detailed critique of Morito's work, see Smith 2008b.

15. While environmentalists have identified many sources of such potentially destructive alienation from nature—from philosophical systems like Cartesian dualism to the capitalist labor process—some form of alienation, in the sense of a recognition of varying patterns of difference/distance among the self, other humans, and more-than-human others is an important aspect of the very possibility of ethics.

16. While several authors try to enlarge the scope of a Levinasian difference ethics to animals (see Aaltola 2002), very few try to do the same ecologically. The key text in ecological terms is Llewelyn's (1991a) *The Middle Voice of Ecological Conscience*. This intricate text, now out of print, deserves a more detailed response than is possible here.

17. Alford (2002, 37) contrasts Levinas's otherworldliness with Murdoch who, he says, is "content to remain within the world of beings." Alford's claim could hardly be further from Diehm's reduction of face to body (see earlier in this chapter), proving, if nothing else, Alford's point about the "Levinas effect": "the ability of Emmanuel Levinas's texts to say anything the reader wants to hear" (24).

18. In fact, although Murdoch wrote one of the first philosophical books on Sartre's work in English (Murdoch 1987 [1953]), she was actually very critical of the level of abstraction that characterizes Sartre's humanism both in terms of its early emphasis on *absolutely* free choice, the supposedly unlimited ability of humans to "leap out of their surroundings" (1987, 21), and his later Hegelian/Marxist

turn where each individual is supposed to be knowable as a *"universal singular* totalized and universalized by his epoch, which he retotalises and reproduces in himself" (22). Despite the radical differences between these positions, they both insist on subsuming the complicated reality of individuals and their ethical choices under abstract concepts (60).

19. Arendt (2005, 7) remarks that "persuasion is a very weak and inadequate translation of the ancient *peithein,* the political importance of which is indicated by the fact that Pēitho, the goddess of persuasion, had a temple in Athens."

20. This is not to say that Plato is, in any sense, central to Arendt's detailed and historically particular understanding of totalitarianism as a specifically *modern* social form, but that Plato is the first philosopher to reduce political action to "giving and executing orders" (Arendt 1975, 325).

21. This is also why Vattimo states, "I sometimes call myself an anarchist. I have proposed . . . that we take seriously the idea from a book by Reiner Schürmann [who] . . . emphasized how Heidegger had preached the end of the epoch dominated by an archē, by the principle, so we now live in an anarchic age" (Caputo and Vattimo 2007, 113). See Schürmann 1990.

22. This, of course, raises questions about Murdoch's early adherence to the Communist Party and its systematic attempts to eradicate plurality, political difference, and people in the name of political equality and historical necessity (Lefort 2007).

23. Although, as Moyn (2005, 116) points out, Barth rejected the term *negative theology* where his own work was concerned, it explicitly emerged from this tradition. Barth's influence can, Moyn claims, also be seen in terms of the adoption of the "talismanic phrase 'the other'" and his radical break with secularism in terms of the role of revelation. Moyn's historically informed discussion of Levinas's sources tends to play down some previous claims concerning the formative intellectual influence of Jewish theologian Franz Rosenzweig on Levinas.

24. "When Levinas later tried to make his own philosophy of intersubjective transcendence purely secular, he did so, it bears noting, in spite of the implication of his own argument from the 1930s that an intersubjective definition of transcendence might remain cryptotheological rather than secular, ultimately dependent on the relation between God and man that it tried to cast in purely human terms. Secularization must involve an operation that ends by truly surpassing the theological point of view" (Moyn 2005, 186). Of course, we might question (following Vattimo) whether secularization "ends" anywhere in particular.

25. "I would like to bring the wide world to you this time. I've begun so late, really only in recent years, to truly love the world. . . . Out of gratitude, I want to call my book on political theories 'Amor Mundi'" (Letter to Jaspers in Young-Bruehl 2006, 79). The book, as Young-Bruehl notes, became *The Human Condition.*

26. For Murdoch, too, ethics emerges through facing the reality of our individual mortality, the inevitability of the limiting event that will ultimately erase all our self-interested possibilities, that "acceptance of our own nothingness which is an automatic spur to our concern with what is not ourselves"(1970, 103).

"A genuine sense of mortality enables us to see virtue as the only thing of worth; and it is impossible to limit and foresee the ways in which it will be required of us" (99).

3. Primitivism

1. The continued underlying presence of myth within the discourses of modernity is a theme detailed at length in Adorno and Horkheimer's *Dialectic of Reason* (1973), which begins with the apposite declaration: "In the most general sense of progressive thought, the Enlightenment has always aimed at liberating men from fear and establishing their sovereignty. Yet the fully enlightened earth radiates disaster triumphant."

2. In other writings, Rousseau's account seems much closer to that of Hobbes. Rousseau sometimes suggests it was only the fact that there was little social interaction between widely separated individuals that stopped the war of one against all. "These barbaric times were a golden age, not because men were united, but because they were separated. . . . His needs, far from drawing him closer to his fellows, drove him from them. If you wish men would attack each other when they met, but they rarely met. A state of war prevailed universally, and the entire earth was at peace" (1986b, 33).

3. Of course, there are many other elements. For example, the roles that the emerging sciences play in the creation of early modernism's political mythology vary considerably. On the one hand, they are agents of secularization; on the other hand, they provide new raw material for the anthropological machine. Clearly, the anthropological interests sparked through global encounters with other peoples are significant. That said, it might be argued that Locke's being influenced by prevailing scientific attitudes and even having contributed to the "evolution of a mechanistic materialism" (Gare 1993) and the "cultural assimilation of Newtonianism" (Mathews 1994, 20) play no significant role in his political archeology. The case is quite different where the mechanistic materialism of Hobbes is concerned.

4. The term *primitivism* actually connotes a heterogeneous community of writers, theorists, and activists, some of whom would regard the term as unnecessarily limiting. For example, in 2001 the UK edition of *Green Anarchist* magazine, long the main proponents of primitivism in Britain, dropped the phrase "for the destruction of civilization" from its masthead, announcing that it was "sloughing off the millstone of primitivism" (Anon., 2001, 8).

5. Moore, John, 2001, "A Primitivist Primer," http://www.eco-action.org/dt/primer.html.

6. In Zerzan's (2005) words, "We should avoid idealizing pre-history, refrain from positing it as a state of perfection. On the other hand, hunter-gatherer life seems to have been marked, in general, by the longest and most successful adaptation to nature ever achieved by humans, a high degree of gender equality, an absence of organized violence, significant leisure time, an egalitarian ethos of sharing, and a disease-free robusticity."

7. Anon., 1995, "Primitivism: Back to Basics," *Green Anarchist* 38, http://the anarchistlibrary.org/HTML/Anonymous_Primitivism_Back_to_Basics_.html.

8. See Bookchin, M., n.d., "Wither Anarchism? A Reply to Recent Anarchist Critics," http://dwardmac.pitzer.edu/Anarchist_Archives/bookchin/whither.html. Bookchin regarded this situation as indicative of an egoistic and narcissistic turn away from notions of social revolution. The epithet "lifestyle," used by Bookchin in an entirely derogatory manner, fails, however, as Mark Smith (1998) has noted, to recognize the importance of new patterns of social resistance associated with the activist politics of New Social Movements and intentional communities (Melucci 1989). The "expressive identities" Hetherington (1998) associated with these social movements have little in common with that egoism which is, ironically, histori-cally more closely associated with dominant strands of Enlightenment humanism.

9. Bookchin rarely let complex realities stand in the way of his opinions of others' work. In fact, Watson (2004, 36) has been quite critical of some aspects of primitivism, explicitly rejecting the kind of fundamentalist version of primitivism that leads some to think the only solution to our current predicament is to try to return to a forager existence. This kind of primitivism certainly does not provide a practical politics for any but a tiny minority of people and, like all fundamental-isms, buries the subtleties of human existence under an overly simplistic ideology, a "reductionist legend in which primordial paradise is undermined by an ur-act of domestication so far back in time that one may as well give up speech, abandon the garden, and roll over and die."

10. Perlman (1983, 7), for example, refers to contemporary peoples such as the !Kung and explicitly claims that "the state of nature is the community of freedoms. Such was the environment of the first human communities, and such it remained for thousands of generations." Yet his essay also adopts a deliberately mythic form that resists attempts to read it as the "straightforward" kind of history it sets itself against.

11. "Feuerbach resolves the religious essence into the *human* essence. But the human essence is no abstraction inherent in each single individual. In its reality it is the ensemble of social relations" (Marx 1975a, 423). Understanding the speci-ficity of human labor in terms of social relations is clearly secularizing and theoreti-cally vital. (Although, insofar as it remains within an anthropological framework, it just pushes the defining question concerning human origins back to what does and does not count as properly social.) But then this understanding should mean that *all* claims to ownership of or authority over nature based in the admixture of human labor should be recognized as historically specific and socially mediated. That is, there is nothing about labor per se that can justify human proprietor-ship over the nonhuman world. Yet this is not Marx's view, and labor still retains something of a mystical aura as a theoretically transcendent category defining the (social) essence of humanity.

The early Marx does recognize the impossibility of *pinpointing* human origins. "Who begot the first man, and nature in general? I can only answer: Your ques-tion is itself a product of abstraction. . . . But since for socialist man the *whole of what is called world history* is nothing more than the creation of man through human labour, and the development of nature for man, he therefore has palpable and incontrovertible proof of his *self-mediated* birth, of his *process of emergence*" (1975b, 357).

12. Radical ecologists have long recognized Marxism's incompatibility with an environmental ethics. In Val Routley's (Plumwood) words, Marx "continues to laud the *objectification* of nature and its reduction to the status of a mere utility. . . . Nature is apparently to be respected to the extent, and *only* to the extent, that it becomes man's handiwork, his or her artifact and self-expression. . . . At the same time the respect position regarding nature which is a feature of the thought of many 'primitive' people is dismissed, in good Enlightenment style, as mere 'prejudice' and superstition, a stage which must be overcome in order to realize full humanity. The view of nature as man's body seems to carry also the unattractive implication that nature is man's *property*" (1981, 243; see also Eckersley 1992, chap. 4).

13. Baudrillard (1992, 105) suggests that productivist Marxism fails to adequately critique dominant labor theories of value. "Radical in its *logical* analysis of capital, Marxist theory nonetheless maintains an *anthropological* consensus with the options of Western rationalism in its definitive form acquired in eighteenth-century bourgeois thought."

14. And not just anarchists! Again, it is easy to dismiss primitivists' simplistic, all-or-nothing arguments concerning the relationship between civilization and civility—in the sense of socially imposed norms of behavior, of morals and manners. However, even extremely sophisticated sociological accounts, like Norbert Elias's *The Civilizing Process,* share some elements of this analysis (Smith and Davidson 2008). Elias (1982, xii) charts what he calls "the psychical process of civilization" in terms of both a history of evolving manners and the relationship between this process and centralized state formation in Western Europe. This psychical process is not reducible to individual psychology but is intended to reveal underlying patterns, movements, and trends of social expression and repression as they develop in and through the material processes of everyday life. While Elias attempts to provide a nonjudgmental sociological account of the figurational dynamics of the civilizing process, it is nonetheless primarily envisaged as a unidirectional trend. This trend is, however, by no means presented as a celebratory account of the "rise" of civilization but pays special attention to manners' and morals' potentially repressive as well as liberative influences. For Elias, the civilizing process is marked by an advancing threshold of embarrassment, an increasing delicacy about, and self-regulation of, bodily processes in public places. Drawing on etiquette manuals and other historical sources, he traces the ways in which, for example, it gradually becomes regarded as good manners that individuals should use a handkerchief rather than hand or sleeve to blow their nose and as courteous to refrain from scratching or spitting in the street rather than just over the dinner table! The anarchic associations of punk rock entirely reversed this norm in Britain, but whether the showers of spit aimed at bands were liberating or simply

messy, unhealthy, and disgusting is a moot point. Of course, taken literally, there would be no punk rock, no streets, and no dinner tables in a primitivist utopia.

15. Turnbull's description of the Ik, a people removed from their traditional gatherer-hunter lands by the creation of a national park, famously described a "society" without any remaining moral norms, one characterized only by a constant struggle for individual survival. In a strange inversion of Hobbes, it was the Ik's experience of the destructive forces of modernization (in the counterintuitive form of conservation management, not industrialization) that threw them into this unfortunate condition. Turnbull (1972, 31) presents the Ik's situation as a moral allegory of modernization's effects from which civilized people should learn about the social fragility of human goodness while also insisting that it "is a mistake to think of small-scale societies as 'primitive' or 'simple.'"

16. *Acéphale* was the title of the journal and secret society initiated by Bataille in 1936, a symbolic figure pictured by Masson, as a "headless figure neither man nor god, his feet firmly rooted on the earth, a deaths head in place of his sex, a labyrinth in his belly, stars on his chest, a knife in one hand, a flaming heart in the other Acéphale" (Kendall 2007, 129–30).

17. This, as is today the case with primitivism, is not to say Bataille was apolitical in terms of his contemporary engagements, as shown by his close prewar involvement in organizing Counter-Attack—hostile to every form of institutional authority, but especially fascism (Kendall 2007, 126). It is to make a point about the disjunction between ethical and political means and apolitical and amoral ends and how the desire for the latter compromises the initiating freedoms of the former.

18. The view of ethics as nothing more than social repression can only regard stronger moral prohibitions as indicative of stronger social repressions, and so this simplistic understanding of ethics easily turns from resistance to oppression to ever-more radical forms of evil as the "liberated" individual tries to root out even the most fundamental of concerns for others' existence. It is not accidental that the idea of human sacrifice, and Bataille's purported offering of himself as such a sacrifice, was one of his and Acéphale's obsessions.

19. Moral absolutism in the sense intended here does not necessarily imply a "'Ten Commandments' idea of morality" (Honderich 1995, 2), a specific list of prohibitions that are made obligatory or placed beyond political debate. Rather, it refers to the claim that the development and defense of an ethical life is only guaranteed by the existence of a political regime that, when necessary, wields absolute authority. In other words, it seeks to reduce ethics to the imposition of a moral order.

20. In addition to the verdicts guilty and not guilty, Scottish law employs "not proven," which leaves a continuing taint of suspicion hanging over the accused. This morally indeterminate situation is perhaps closer to that of everyday life in a surveillance society.

21. Of course, language can be used ideologically, just as it can and does operate to symbolically divide the world into (more or less) fixed categories. But even in saying this, Zerzan proves, and knows that he proves, that it also carries within

it the critical potential of a mode of individualizing expression and a way of relating to others that would otherwise be entirely unavailable to him. As already remarked, Levinas (2004) pays special (ethical) attention to this difference between saying and said, between that which initiates and expresses the freeing play of difference through language and that which conceptually fixes meaning in language, but he does not fall into the all-or-nothing trap of thinking that one can have *absolute* freedom within (or without) language. Zerzan (1999, 40) also claims that "as a formalizing, indoctrinating device, the dramatic power of art fulfilled this [upper-Paleolithic] need for cultural coherence and the continuity of authority." In this way, every form of negativity with no use, that is, any initiating possibilities that might be open to human individuals but are not potentially shared with all of nature, are simply rejected as residually alienating aspects of civilization.

22. Even though Arendt wants to define particularly human forms of natality, of action and speech, she also recognizes that "this character of startling unexpectedness is inherent in all beginnings and all origins" (1958, 178), even the infinite improbability of the origin of life from inorganic matter and "the evolution of human life out of animal life" (178).

23. There are many ways in which this ongoing creative vitality might be approached (though not captured) by thought, and some of these also challenge the metaphysical bounds of the anthropological machine. One possibility might be to revisit explicitly vitalist philosophies like Bergson's, perhaps especially in terms of Deleuze's (1997) interpretation of the *élan vital* as an expressive differential movement informing the world rather than a mysterious originating principle. Another example, although somewhat limited by its debts to Marxist forms of productivism (see earlier discussion), is Lefebvre's (1994) distinction between differences that are *induced* (minimal)—that are included within and usually collude with a system (for example, the differences between 2, 4, 6, 8, and so on, in a numerical order)—and differences that are *produced* (maximal)—that are excluded by and challenge the idea of absolute regularity, continual process, universal laws, and so on. The unpredictable variations and events that provision evolutionary change exemplify this notion of maximal (productive) difference (see Smith 2001d; 2002a).

24. Another of Camus's remarks might be pertinent here. If the future is divested of any meaning, then "we have nothing to lose except everything. So let's go ahead. This is the wager of our generation. If we are to fail, it is better, in any case, to have stood on the side of those who choose life than on the side of those who are destroying" (Camus 1961, 246).

25. *Ecological anarchy* is intended to refer to primitivists *and* anarchists sympathetic to aspects of primitivism's analysis who nonetheless reject their all-or-nothing principle, for example, Watson (1996; see earlier discussion).

4. Suspended Animation

1. *Lifeless* here refers to a situation in which, even if something is recognized as a living entity, a laboratory rat for instance, that fact is granted no evaluative

importance in and of itself; its life can be terminated ("sacrificed" in the terminology of the laboratory) whenever the experimental protocol determines it necessary. The rat is to be regarded with "detached objectivity" and therefore as nothing more than a detached "object." Evernden (1999, 14) refers to this replacement of living beings with objective abstractions as "cutting the vocal chords" of the world.

2. Nor, after this initial and disastrous intervention in politics, did he attempt to explicitly engage in theorizing politics or ethics as such.

3. This is not to say that nature is regarded as bare life, since this is a politically determined human condition, a denial of what are presumed to be human possibilities, but as a standing reserve, a systematically orderable resource.

4. Agamben makes a distinction between constituent power, which he associates with political constitutions that concentrate sovereign power in states, and constitutive power (potential), the power to create political communities.

5. Sandilands adds that the term *political animal* is a "phrase that recognizes that we are constituted simultaneously as creatures of [political] discourse, as creatures that discourse can never entirely describe, and as the paradox itself" (1999, 206).

6. This, it should be noted, is at odds with any reductivist notions that human politics are biologically constituted or that nature is just a social construct.

7. As Calarco (2008) argues, much of Agamben's early work "contains all the dogmatic elements . . . seen in Heidegger's and Levinas's discourses on animals: a simple human–animal distinction; lack of attention to existing empirical knowledge about animals . . . and the invariant concern to determine the supposedly unique human relation to the world and history—as if what constitutes the ground of a supposedly unique mode of human existence is the sole (or even primary) thing at stake for philosophical thought."

8. Although Agamben does sometimes speak of the possibility that "man himself will be *reconciled* with his animal nature" (Agamben 2004, 3, my emphasis)—a much more hopeful image.

9. See also, for example, Zimmerman's (1994, 376) claim that "unless contemporary people act to alleviate widespread social oppression, to halt senseless ecological destruction, and to articulate creative and critical responses to oncoming technological developments, the human phase of evolutionary history may soon come to an end."

10. In *The Open,* a quasi-Hegelian motif reasserts itself because Agamben seems to suggest that the separation of human and animal will relieve the latter from its obscure expectation of "revelation," a revelation open only to humanity; this seems to suggest a phylogenetic telos toward the evolution of humankind.

11. Heidegger (1993b) describes this relation in terms of the relation between Earth and World. See also Haar (1993).

12. Bartelson (1995) provides a detailed genealogy of sovereignty, which, rather than trying to pinpoint its exact origins, traces its conceptual antecedents back beyond the first apparent use of the term in the thirteenth century (88) and also details its later connections with the state form.

13. Pateman (1985, 168) further argues that "to take self-assumed obligation [*responsibility* would be a better term] seriously as a political ideal is to deny that the authority of the liberal democratic state and the (hypothesized) obligations of its citizens can be justified."

14. While biopolitics is, for Foucault, a defining feature of modernity, Agamben associates it with *all* claims to sovereign power tracing it back to Roman Law—which is where the term *Homo sacer*, which he later equates with Benjamin's notion of bare life, originates. *"It can even be said that the production of a biopolitical body is the original activity of sovereign power"* (1998, 6). Today, Agamben argues, this management of bare life has become a general task placing entire populations in a zone of indistinction, "exposed and threatened on the threshold in which life and law, outside and inside, become indistinguishable" (28).

15. And, of course, Agamben (2004, 22) argues that the anthropological machine played its part here too: The concentration camp can itself be understood as "an extreme and monstrous attempt to decide between the human and the inhuman, which has ended up dragging the very possibility of the distinction to its ruin."

16. Latour distinguishes between "Science" (capital *S*), which serves to maintain that unified conception of nature operating as part of the modern political settlement, and the "sciences" as diverse practices of knowledge production that are a key element in the composition of any community.

5. Risks, Responsibilities, and Side Effects

1. Sandilands (1999, chap. 7) presents a detailed ecofeminist reading of Arendt's work in relation to the work of Nancy Fraser, Drucilla Cornell, Judith Butler, and Jacques Lacan. In particular, she examines the implications of the gendered assumptions underlying Arendt's distinction between the private and public (political) spheres.

2. For a critical review of Blühdorn's work, see Smith 2002b.

3. This includes using Science to such purposes—where Science is defined by Latour (2004, 10) as "the politicization of the sciences through epistemology in order to render ordinary political life impotent through the threat of an incontestable nature."

4. Although Beck is at pains to point out that the self that results from what he terms the process of individualization in risk society is by no means the same as the isolated, rational, calculating Homo economicus so closely associated with modernist political theory.

5. Latour's (2004, 226) lack of respect for the philosophy of political ecology is, he claims, due to environmentalists not making use of the resources available to them in the "philosophy of science" and "comparative anthropology," that is, in his own fields of interest! Is this parochial of them or him? But then, he seems to be candid, to have read little or nothing written by ecologically concerned theorists that might have contradicted this preconceived caricature.

6. Again, it is important to reiterate that not only humans and not all humans have this character of being a being who cares and for whom ethics is a possibility.

6. Articulating Ecological Ethics and Politics

1. The "'virtue' of integrity as a standard is that it is impartial between competing conceptions of the political good" (Westra 1994, 200, quoting Beitz).

2. Aristotle was, after all, very selective about just which humans he thought capable of living a political life.

3. Although Levinas too sometimes emphasizes the public aspects of the ethical relation that composes the between us: "Everything that takes place here 'between us' concerns everyone, the face that looks at it places itself in the full light of the public order" (1991, 212).

4. Arendt emphasizes politics as *the* key active dimension of freedom and individuation, but following Bataille and others, we have already argued that there are many other dimensions that constitute such excessive possibilities, such as art, love, play, and wildness (see chapter 7).

5. When Arendt and Levinas are mentioned together, it is usually in respect of a third figure, for example, Simone Weil (Hermson 1999) or St. Augustine (Astell 2006). An exception is Anderson (2006), but the negative parody of Arendt's work he presents severely restricts its value. By contrast, Schmeiden (2005), who focuses on the question of "productive work," is highly critical of Levinas's "a-symmetrical ethical monotheism," which he contrasts with Arendt's "symmetrical and polytheistic intersubjectivity" (223).

6. Although largely taken up, and for good reason, by the nondogmatic political left, this ambiguity is often remarked on, and Levinas has even been associated with the neoconservative right through the figure of Leo Strauss (see Batnitzky 2006). Batnitzky, though, emphasizes the primacy of Levinas's prophetic and messianic writings on Judaism, arguing that Levinas can be read as a religious rather than a political Zionist and that this renders his secular "political thought meaningless"; it results in a "nonpolitical politics" (160). This seems completely at odds with Critchley (2002a, 22–24), who claims "one of the prevailing and potentially misleading assumptions about Levinas's work is that he is a Jewish philosopher"; rather, he is a "philosopher and a Jew" and one whose "attempt to traverse the passage from ethics to politics" is both insistent and coherent.

7. Interestingly, Critchley's (2007) more recent work has developed a specifically anarchistic (in the more usual political sense) reading of Levinas (see later discussion). It will be interesting to see whether and how this might be developed along Arendtian lines.

8. In this particular interview, Levinas was speaking specifically about the state of Israel in the context of the September 1982 massacre of several hundred Palestinians in the Sabra and Chatila refugee camps. Levinas was asked, "And for the Israeli, isn't the 'other' above all Palestinian?" (Finkeilkraut in Levinas 1989, 294). His answer was, at best, equivocal. It also seems necessary to ask whether

such an ethical necessity would accrue to the defense of all states or whether it is tied specifically to Levinas's understanding of the state of Israel as a "coincidence of the spiritual and political" (Levinas in Shapiro 1999, 68). If the latter, then this is, as Shapiro (1999, 68) notes, astoundingly partisan. If the former, it would seem to raise many more ethical and political questions.

9. As Levinas (1998, 159) remarks, it "is not then without importance to know if the egalitarian and just State in which man is fulfilled . . . proceeds from a war of all against all, and if it can do without friendship and faces" (see also Critchley 1999, 221). But it is also important to know what the insistence that ethics is *first* philosophy really amounts to if the existence of a political state and its (Hobbesian) role in the struggle for existence is given priority over ethical concerns for the lives of others outside of that state apparatus.

10. *Outwith,* a term that is part of the Scottish vernacular, has exactly the connotations needed here, suggesting both proximity (a being-with) and excess (an overflowing that cannot be contained within and yet is not just exterior to and separated from that being discussed).

11. Although making no reference to Levinas, Hailwood (2004) also frames his argument explicitly in a language that recognizes the otherness of nature.

12. And some, like Eckersley (2004, 244), approach this largely in terms of a *constitutionally* defined politics, envisaging, among other measures, the "constitutional entrenchment of an independent public authority—such as an environmental defenders office—charged with the responsibility of politically and legally representing public environmental interests, including the interests of non-human species and future generations."

13. In a letter to Ernst Schoen, Benjamin (1994, 138) recalls how he suggested that "it was a mistake to postulate anywhere a purity that exists in and of itself and needs only to be preserved. This tenet seems to me to be important. . . . The purity of an essence is *never* unconditional or absolute; it is always subject to a condition. This condition varies according to the essence whose purity is at issue; but this condition *never* inheres in the being itself. In other words the purity of every finite essence is not dependent upon itself. The two essences to which we primarily attribute purity are nature and children. For nature, human language is the extrinsic condition of its purity." This passage is also cited in a slightly different translation by Agamben (2005, 61).

14. Althusser distinguishes between specific ideologies and ideology in general, the representations provided by the symbolic order of society within which nascent individuals are insinuated as subjects through the process of interpellation (Smith 2001a, 177).

15. For an anarchist critique of the concept of hegemony, see Day 2005. Unfortunately, as the history of anarchy shows, it is all too easy to succumb to such pressures, for example, by invoking oxymoronic anarchist principles or reimporting the anthropological machine in the guise of an alternative metaphysics of a (beneficent) ideal of human nature. Atheists too have often simply replaced faith in God with an anthropological faith in a metaphysical and logocentric prin-

ciple of human reason. The ridiculous spectacle of archrationalist and atheist Anthony Flew's recent recantation of his faith in reason and his docile return to the Christian fold merely illustrates the contiguity between these foundational forms. The ways in which many postwar European intellectuals simply switched allegiances back and forth between an unshakable faith in Stalin, the infallibility of the Pope, or the thoughtless advocacy of global capitalism might provide other examples which also illustrate the difficulties, but also the ethicopolitical value, of sustaining (affirming) weak thought.

16. In this way, too, one might also speak, as Benjamin (1998) did, of "language as such," that language which exceeds human language, that is the communicable expression of beings. For the ecological implications of this view see Smith 2001b.

17. Lacan's theses also suggest yet another reason, in this case psychological rather than sociological or ethical (like Levinas and Murdoch), why the individual is never sovereign and self-contained. In Lacanian terms, the human subject's experience of lack and subsequent desire for completion is ultimately unachievable precisely because this subject is originally defined over and against what it is not, what is Other. To change the relation to the Other, to try to own it in some way, inevitably changes and redefines rather than completes the self. It thereby reveals to the human subject the dizzying absence of any essential fixed self-identity, any *final answer* to the question of who they really are.

18. But see Lefort's essay "The Permanence of the Theological-Political," in which he states that modern democracy is "the only regime to indicate the gap between the symbolic and the real by using the notion of a power which no one—no prince and no minority—can seize. It has the virtue of relating society to the experience of its institution" (1988, 228). As it stands, this remark unfortunately seems to leave open the question of whether this power can be "seized" by the majority, or those claiming to represent them, not to mention the very differing experiences of this "institution."

19. This political use of Lacan might be contrasted with the neo-Leninism of Slavoj Žižek (2001, 3), who argues that the widespread academic acceptance of Arendt's work and even the attempt to contrast democracy with totalitarianism is somehow indicative of the theoretical failure of "the Left." Žižek also discusses Derrida's, Levinas's, and Critchley's understanding of the relation of ethics to politics. In one of his later works, Žižek (2008, 441) suggests that ecology is "the new opium of the masses" but then proposes using the ecological challenge to support what seems suspiciously like a new form of totalitarianism, one that suppresses "liberal freedoms," guarantees conformity to norms through terror and ruthless punishment, advocates technological control of "prospective" lawbreakers, and makes informers into heroes (461). If this Orwellian vision really is indicative of what Žižek regards as the Left's true vocation, there is little to regret about its failure. His almost complete ignorance about the ecological issues on which he pontificates is quite astounding. At one point, he even cites the novelist Michael Crighton as an authority! Crighton's novel *State of Fear* has as its hero a scientist who discovers that climate change is a hoax, a view explicitly shared by

Republican Senator James Inhofe who, on this scientifically worthless basis, summoned Crighton to testify before the Senate in 2005.

20. This recognition evolved through his ongoing dialogue with the work of anthropologist Pierre Clastres (see Lefort 2000). Clastres's work, especially his "Society Against the State" (1987), a short section of which is included in Zerzan (2005), has been influential in anarcho-primitivist evocations of a state of nature. Lefort (2000, 216), however, regards this as misreading Clastres's intentions. "One cannot reasonably attribute to Clastres the thesis that primitive society would offer us the image of the 'good society,' that it might furnish us with a model to which our contemporaries ought to become attached in order to deliver themselves from the perversity of modern institutions."

21. "Lefort's thought is far from any form of anarchism" (Flynn 2005, xxv).

22. Hardt and Negri's understanding of biopolitics is not the same as Agamben's— it refers to political aspects of questions of life *(bios)* rather than regarding it as a form of governance associated with the reduction of politics to matters of survival. As Campbell (in Esposito 2008, viii) notes: "For Antonio Negri, writing with Michael Hardt, biopolitics takes on a distinctively positive tonality when thought together with the multitude." Greenpeace's campaigns are thus biopolitical for Hardt and Negri (2004, 282), whereas as far as I am concerned (assuming the ecological twist given to Agamben's approach) they are explicitly antibiopolitical, they are about setting whales free from the principle of human sovereignty (see earlier discussion).

23. As Paterson (2000, 62) argues, "Whether or not we term the result 'anarchist' . . . the dominant political prescription within Green Politics is for a great deal of decentralization of political power to communities much smaller in scale than nation-states, and for those communities to be embedded in networks of communication and obligation across the globe (that is, for them to be non-sovereign)."

7. Against Ecological Sovereignty

1. The provisions of the treaties associated with the Peace of Westphalia, which signaled the end of the Thirty Years War in 1648, are generally regarded as the first expression of the principles supposed to legitimate the sovereign authority of modern nation-states.

2. References to political solutions in Gaian literature are scarce and rarely engage explicitly with sovereignty because Gaia's advocates have been primarily interested in its spiritual and/or scientific repercussions to the exclusion of all else (Midgley 2006). These concerns tend to find expression in their preferences for individualistic lifestyle and/or technical fixes to ecological problems. Even where globalization is (at least implicitly) addressed, no clear model of political authority emerges. For example, Stephan Harding's (2006, 237) recent *Animate Earth* regards global institutions like the World Bank and multinational corporations as "major instruments of the war against nature," advocating instead a "global steady-state economy by means of worldwide legal enforcement" (238). But what

this actually entails politically is never explained, since Harding claims "the real change has to be an inner one" (239). Cullinan (2007, 33) too suggests a "need for legal and political systems that reflect this Gaian perspective" but has little to say about the actual forms taken by the "governance systems" he envisages will "regulate humans" except that any socioculturally specific varieties of Earth jurisprudence "must be shaped to accord with the natural laws that govern the system as a whole" (34).

3. Comparisons might be made here with the technologically and biologically based antidemocratic solutions of the "Open Conspiracy" of intellectuals proposed by H. G. Wells in the 1920s, later reflected in Aldous Huxley's (1971 [1932]) *Brave New World*.

4. Of course, a few noncompliant states still exist, but their retention of territorial prerogatives over the economy is usually a sign of their being both isolationist and totalitarian. These examples of exercising state sovereignty do not provide good models for any ecological politics.

5. Despite Eckersley's (2004, 208) promise to provide an account of the organizing principle of sovereignty, her protean conception does not allow her to do so. This is also true of other writers in this field, like Litfin (see Eckersley 2004, 209), who argue that it is more important to understand sovereignty in terms of its variable operational norms rather than juridical–philosophical principles.

6. No one would dispute that the claim that something is "a matter of political sovereignty" *can* be deployed as if it were a principle that justified excluding external political or economic interference in a nation's practices or, for example, nationalizing key industries within its territorial jurisdiction. In such cases, sovereignty is supposed to represent (in a way that should, according to Westphalian presumptions, place it beyond international criticism) the political right of a state to self-manage its internal affairs and ensure its own survival as an entity. The problem here is not just the hypocrisy of the rhetoric of sovereignty—since such actions are often precisely those deemed to be sufficient grounds for the externally orchestrated overthrow of governments by other more powerful sovereign states— but the less obvious (Schmittian) point that sovereign power is defined entirely in terms of having the *ability* to make such decisions. It has nothing whatsoever to do with the political form of the state per se, which can be more or less democratic, royalist, communist, feudal, totalitarian, and so on—that can, in other words, encompass situations with more or less (ecological) politics as such. So either the appeal to sovereignty is little more than an appeal to nationalism (as it usually is with the conservative) or/and it is an expression of having (or wanting to have) the power to exercise such a decision. In neither sense is sovereignty something necessarily, primarily—or perhaps at all—associated with guaranteeing politics as such. Greening democratic politics is therefore a very different project from greening sovereignty. One can, of course, concoct arguments to the effect that only a democratic polity might have a claim to legitimately exercise this principle of sovereignty. But in doing so, one not only reiterates a colonial occidental political bias (and, after all, which Western nation is really democratic?), one either ignores

the dangers inherent in the ultimate expression of sovereignty, which is precisely to be able to internally suspend (democratic) politics, or one undercuts the very principle of sovereignty itself insofar as one makes it entirely subject to and dependent on the presence and exercise of democratic politics. Either way, the emphasis should be on greening politics, not sovereignty.

7. Think of Woody Allen's film *Annie Hall* in which the young Alvy Singer is taken by his mother to see the doctor for depression. "It's something he read," she says. "What is it, Alvy?" "Well, the universe is everything, and if it's expanding then some day it will break apart and that will be the end of everything." "What, is that your business?" says his mother. "He's stopped doing his homework." "What's the point?" says Alvy. To which she retorts, "What has the universe got to do with it? You're here in Brooklyn and Brooklyn is *not* expanding!"

8. Indeed, Agamben (1998, 113) sometimes explicitly relates his work to Nancy.

9. As Nancy (2007, 51) explains, that the world is created ex nihilo does "not mean fabricated with nothing by a particularly ingenious producer. It means instead that it is not fabricated, produced by no producer, and not even coming out of nothing (like a miraculous apparition), but . . . nothing growing *[croissant]* as *something*. . . . In creation, a growth grows from nothing and this nothing takes care of itself, cultivates its growth. The *ex nihilo* is the genuine formulation of a radical materialism, that is to say, precisely without roots."

10. Nancy (2007, 94) explicitly criticizes the view that modern sovereignty is a secularized theological concept, as Schmitt and Agamben maintain (see earlier discussion), but I find his arguments unconvincing.

11. Perhaps due to Bataille's influence, Nancy recuperates the concept of sovereignty in a specific way relating to the (excessive) creative abilities of beings and their capacities to resist incorporation in any totalizing system of representation. So understood, "Sovereignty is singularity in its very specific resistance to political appropriations, not the exalted, superlative power of legitimacy that overrides this resistance" (Hutchens 2005, 168). Nancy's "sovereignty" is not beholden to any principle; one might say it is a matter of authorship rather than authority, and of the "reclamation of sovereignty at its roots, which is *nothing*" (Nancy 2008, 105). But this does not alter the fact that political sovereignty remains a principle that is always invoked to overcome (that which is deemed excessive) resistance. Unfortunately (as Agamben argues concerning Bataille, see earlier discussion), Nancy's recuperation broadens the notion of sovereignty in a way that loses its political specificity. It fails to distinguish between the many dimensions of creative excess (of freedoms) and the excessive political power that is free to decide to restrict some or all of those freedoms as and when it chooses. So when Nancy (2007, 109) asks: "And what if sovereignty was the revolt of the people?" one might reply that the *revolt* of people (singular *and* plural) is always against a principle of sovereignty. We might also recall that, as Lingis (1997, 211) remarks, "It is not the abundance of commodities available for consumption that make free individuals possible; instead, the value assigned to the free individual, as a positive force that freely produces his or her own existence by

affirming the sovereignty of his or her desire, produces the contemporary form of consumption."

12. Here again, a provisional ecopolitics must differ with Nancy's formulations. Nancy (2008) refers explicitly to sharing the world as a form of justice, but this is both indicative of his residual emphasis on the shared *human* world (and consequently his rather uncritical extension of ethicopolitical concepts to encompass the more-than-human world) and the loss of any specific articulation of justice with ethics and politics. In this sense, the problem mirrors (reflects and inverts) forms of ecological naturalism that espouse a vague form of holism that can only look at the way the more-than-human world is (represented as being) "shared" and say Amen (so be it) to that—as if there were some kind of praiseworthy natural justice in operation in the nonhuman world. But to be (to exist), ethically and politically, is never to simply acquiesce to the way the world is. Justice is not given—it is ethically and politically created.

13. The balance of justice is a weighing of one issue against another, a social trade-off. The balance of nature (if it can even be called that) is very different and takes many forms—the fluctuation of the kestrel's wings in its hovering flight, the circulating continuance of the winds, the ingestion and flux of a body, the codependencies of predators and prey, symbiosis, and so on. Whatever their differences, it should be noted that the outcomes of justice and evolution are in no way reducible to the principles of economic equivalence underlying the balancing of financial accounts. Ecology is not, as it is so often portrayed, nature's economy.

Apologue

1. I am aware that a moral fable carries ethical dangers too for those whose existence is put to exemplary purposes. For a detailed discussion, see Smith 2005a.

2. Although seven nations still lay territorial claims to sovereignty over parts of Antarctica (Joyner 1998).

BIBLIOGRAPHY

Aaltola, Elisa. 2002. "'Other Animal Ethics' and the Demand for Difference." *Environmental Values* 11 (2): 193–209.

Abensour, Miguel. 2002. "'Savage Democracy' and the 'Principle of Anarchy.'" *Philosophy and Social Criticism* 28 (6): 703–26.

Abram, David. 1996. *The Spell of the Sensuous: Perception and Language in a More-than Human World.* New York: Vintage Books.

Adams, Carol J. 1990. *The Sexual Politics of Meat: A Feminist-Vegetarian Critical Theory.* Oxford, England: Polity.

Adorno, Theodor, and Max Horkheimer. 1973. *Dialectic of Enlightenment.* London: Verso.

Agamben, Giorgio. 1993. *The Coming Community.* Minneapolis: University of Minnesota Press.

———. 1998. *Homo Sacer: Sovereign Power and Bare Life.* Stanford, Calif.: Stanford University Press.

———. 1999. *Remnants of Auschwitz: The Witness and the Archive.* New York: Zone Books.

———. 2000. *Means without End: Notes on Politics.* Minneapolis: University of Minnesota Press.

———. 2001. *The Coming Community.* Minneapolis: University of Minnesota Press.

———. 2004. *The Open: Man and Animal.* Stanford, Calif.: Stanford University Press.

———. 2005. *State of Exception.* Chicago: University of Chicago Press.

Alford, C. Fred. 2002. "Emmanuel Levinas and Iris Murdoch: Ethics as Exit?" *Philosophy and Literature* 26 (1): 24–42.

Allen, Woody. 1977. *Annie Hall.* United Artists.

Althusser, Louis. 1969. *For Marx.* Harmondsworth, England: Penguin.

———. 1990. *Philosophy and the Spontaneous Philosophy of the Scientists and Other Essays.* London: Verso.

Anderson, Alun. 2009. *After the Ice: Life, Death and Geopolitics in the New Arctic*. New York: HarperCollins.

Anderson, Benedict. 1991. *Imagined Communities: Reflections on the Origins and Spread of Nationalism*. London: Verso.

Anderson, Keith. 2006. "Public Transgressions: Levinas and Arendt." In *Difficult Justice: Commentaries on Levinas and Politics*, ed. Asher Horowitz and Gad Horowitz, 127–47. Toronto: University of Toronto Press.

Anjoulet, Norbert. 2005. *Lascaux: Movement, Space, and Time*. New York: Harry N Abrams.

Anon. 2001. Editorial. *Green Anarchist* 63.

Antonaccio, Maria. 2000. *Picturing the Human: The Moral Thought of Iris Murdoch*. Oxford: Oxford University Press.

Ardilla. 2003. "Intuition as a Crucial Part of Rewilding." *Green Anarchy* 16: 46.

Arendt, Hannah. 1958. *The Human Condition*. Chicago: University of Chicago Press.

———. 1965. *On Revolution*. Harmondsworth, England: Penguin.

———. 1970. *On Violence*. New York: Harcourt Brace & World.

———. 1973. "Civil Disobedience." In *Crisis of the Republic*. San Diego: Harcourt Brace.

———. 1975. *The Origins of Totalitarianism*. New York: Harcourt Brace Jovanovich.

———. 1977. *The Life of the Mind*. San Diego: Harvest.

———. 1982. *Lectures on Kant's Political Philosophy*. Chicago: University of Chicago Press.

———. 1993. *Between Past and Future*. Harmondsworth, England: Penguin.

———. 1994. *Eichmann in Jerusalem: A Report on the Banality of Evil*. New York: Penguin.

———. 2003. *Responsibility and Judgment*. New York: Schocken Books.

———. 2005. *The Promise of Politics*. New York: Schocken Books.

Aristotle. 1988. *The Politics*. Cambridge: Cambridge University Press.

Ascherson, Neil. 2006. "Imagined Soil." *London Review of Books* 28 (7): 11–13.

Assy, Bethania. 2008. *Hannah Arendt—An Ethics of Personal Responsibility*. Frankfurt: Peter Lang.

Astell, Ann W. 2006. Mater-Natality: Augustine, Arendt, and Levinas. *Analecta Husserliana* 89: 373–98.

Attfield, Robin. 1991. *The Ethics of Environmental Concern*. 2nd. ed. Athens: University of Georgia Press.

Bakunin, Michael. 1920. *God and the State*. Indore, India: Modern Publishers, sole agents for Bombay Libertarian Book Dept.

Barnett, Jon. 2001. *The Meaning of Environmental Security*. London: Zed Books.

Barry, John. 1999. *Rethinking Green Politics: Nature, Virtue and Progress*. London: Sage.

Barry, John, and Robyn Eckersely, eds. 2005. *The State and the Global Ecological Crisis*. Cambridge, Mass.: MIT Press.

Bartelson, Jens. 1995. *A Genealogy of Sovereignty*. Cambridge: Cambridge University Press.

Bataille, Georges. 1955. *Prehistoric Painting: Lascaux or the Birth of Art*. Lausanne: Skira.

———. 1991. *The Accursed Share*. Vol. III, *Sovereignty*. New York: Zone Books.

———. 2005. *The Cradle of Humanity: Prehistoric Art and Culture*. New York: Zone Books.

Batnitzky, Leora. 2006. *Leo Strauss and Emmanuel Levinas: Philosophy and the Politics of Revelation*. Cambridge: Cambridge University Press.

Baudrillard, Jean. 1992. *Selected Writings*. Cambridge, England: Polity.

Baxter, Brian. 1999. *Ecologism: An Introduction*. Edinburgh: Edinburgh University Press.

———. 2005. *A Theory of Ecological Justice*. London: Routledge.

Bayley, John. 1999. *Iris: A Memoir of Iris Murdoch*. London: Duckworth.

Beck, Ulrich. 1992. *Risk Society: Towards a New Modernity*. London: Sage.

———. 1997. *The Reinvention of Politics: Rethinking Modernity in the Global Social Order*. Cambridge, England: Polity.

———. 1998. *Democracy without Enemies*. Cambridge, England: Polity.

Beck, Ulrich, and Johannes Willms. 2004. *Conversations with Ulrich Beck*. Oxford, England: Polity.

Bekoff, Marc, and Jessica Pierce. 2009. *Wild Justice: The Moral Lives of Animals*. Chicago: University of Chicago Press.

Benhabib, Seyla. 2006. *Another Cosmopolitanism*. Oxford: Oxford University Press.

Benjamin, Walter. 1973. "Theses on the Philosophy of History." In *Illuminations*. London: Fontana.

———. 1994. *The Correspondence of Walter Benjamin*. Ed. Gershom Scholem and Theodor W. Adorno. Chicago: University of Chicago Press.

———. 1998. "Of Language as Such and the Language of Man." In *One-Way Street and Other Writings*. London: Verso.

———. 1999. *The Arcades Project*. Cambridge, Mass.: Harvard University Press.

———. 2006. "On the Concept of History." In *Walter Benjamin: Selected Writings Volume 4, 1938–1940*. Cambridge, Mass.: Harvard University Press.

Benso, Silvia. 2000. *The Face of Things: A Different Side of Ethics*. Albany: State University of New York Press.

Bergo, Bettina. 2003. *Levinas between Ethics and Politics: For the Beauty That Adorns the Earth*. Pittsburgh, Pa.: Duquense University Press.

Berking, Helmuth. 1996. "Solitary Individualism: The Moral Impact of Cultural Modernisation in Late Modernity." In *Environment and Modernity: Towards a New Ecology*, ed. Scott Lash, Bronislaw Szerszynski, and Brian Wynne. London: Sage.

Bernasconi, Robert. 2005. "Levinas and the Struggle for Existence." In *Addressing Levinas*, ed. Eric Sean Nelson, Antje Kapust, and Kent Still. Evanston, Ill.: Northwestern University Press.

Bernstein, Richard J. 2002. "Evil and Theodicy." In *The Cambridge Companion to Levinas,* ed. Simon Critchley and Robert Bernasconi. Cambridge: Cambridge University Press.

Best, Steven. 1998. "Murray Bookchin's Theory of Social Ecology: An Appraisal of *The Ecology of Freedom.*" *Organization and Environment* 11 (3): 334–53.

Bingham, Nick. 2006. "Bees, Butterflies, and Bacteria: Biotechnology and the Politics of Non-Human Friendship." *Environment and Planning* 38: 483–98.

Biro, Andrew. 2005. *Denaturalizing Ecological Politics: Alienation from Rousseau to the Frankfurt School and Beyond.* Toronto: University of Toronto Press.

Black and Green Network. n.d. *What Is Green Anarchy?* Eugene, Ore.: Green Anarchy.

Black, John. 1970. *The Dominion of Man: The Search for Ecological Responsibility.* Edinburgh: University of Edinburgh Press.

Bloch, Ernst. 1995. *The Principle of Hope.* Vol. 1. Cambridge, Mass.: MIT Press.

———. 1997. *The Principle of Hope.* Volume 3. Cambridge, Mass.: MIT Press.

Blühdorn, Ingolfur. 2000. *Post-Ecologist Politics: Social Theory and the Abdication of the Ecologist Paradigm.* London: Routledge.

Bookchin, M. 1982. *The Ecology of Freedom.* Palo Alto, Calif.: Cheshire Books.

———. 1995a. *Social Anarchism or Lifestyle Anarchism.* Edinburgh: A.K. Press.

———. 1995b. *Re-Enchanting Humanity—A Defence of the Human Spirit against Anti-humanism, Misanthropy, Mysticism and Primitivism.* Montreal: Black Rose Books.

———. 1999. "Wither Anarchism? A Reply to Recent Anarchist Critics." In *Anarchism, Marxism, and the Future of the Left.* San Francisco: A.K. Press.

Bourdieu, Pierre. 1991. *Outline of a Theory of Practice.* Cambridge: Cambridge University Press.

Bruns, Gerald L. 1997. "The Remembrance of Language: An Introduction to Gadamer's Poetics." In *Gadamer on Celan: Who Am I and Who Are You and Other Essays,* ed. Richard Heinemann and Bruce Krajewski. Albany: State University of New York Press.

Butler, Judith. 2004. *Precarious Life: The Powers of Mourning and Violence.* London: Verso.

———. 2007. "'I Merely Belong to Them.' Review of Arendt's *The Jewish Writings.*" *London Review of Books* 29 (9): 26–28.

Calarco, Mathew. 2007. "Jamming the Anthropological Machine." In *Giorgio Agamben: Sovereignty and Life,* ed. Mathew Calarco and Steven DiCaroli. Stanford, Calif.: Stanford University Press.

———. 2008. *Zoographies: The Question of the Animal from Heidegger to Derrida.* New York: Columbia University Press.

Calarco, Mathew, and Steven DiCaroli. 2007. *Giorgio Agamben: Sovereignty and Life.* Stanford, Calif.: Stanford University Press.

Callicott, Baird. 1985. "Intrinsic Value, Quantum Theory, and Environmental Ethics." *Environmental Ethics* 7: 257–75.

Camus, Albert. 1961. *Resistance, Rebellion and Death.* New York: Alfred A Knopf.

——. 1984. *The Rebel.* Harmondsworth, England: Penguin.

Caputo, John D., and Gianni Vattimo. 2007. *After the Death of God.* New York: Columbia University Press.

Castells, Manuel. 1998. *The Rise of the Network Society.* Oxford: Blackwell.

Castree, Noel, and Bruce Braun, eds. 2001. *Social Nature: Theory, Practice, and Politics.* Oxford: Blackwell.

Cavarero, Adriana. 1995. *In Spite of Plato: A Feminist Rewriting of Ancient Philosophy.* Cambridge, England: Polity.

Caygill, Howard. 1997. "The Shared World—Philosophy, Violence, Freedom." In *The Sense of Philosophy: On Jean-Luc Nancy,* ed. Darren Sheppard, Colin Sparks, and Colin Thomas. London: Routledge.

Christoff, Peter. 2005. "Out of Chaos, a Shining Star? Toward a Typology of Green States." In *The State and the Global Ecological Crisis,* eds. John Barry and Robyn Eckersely. Cambridge, Mass.: MIT Press.

Clastres, Pierre. 1987. *Society against the State.* New York: Zone Books.

Coates, Peter. 1998. *Nature: Western Attitudes since Ancient Times.* Cambridge, England: Polity.

Conradi, Peter J. 2001. *Iris Murdoch: A Life.* London: HarperCollins.

Critchley, Simon. 1999. *The Ethics of Deconstruction: Derrida and Levinas.* 2nd. ed. Edinburgh: Edinburgh University Press.

——. 2002a. Introduction. In *The Cambridge Companion to Levinas,* ed. Simon Critchley and Robert Bernasconi. Cambridge: Cambridge University Press.

——. 2002b. Introduction. *Parallax* 8 (3): 1–4.

——. 2007. *Infinitely Demanding: Ethics of Commitment, Politics of Resistance.* London: Verso.

Cronon, William. 1995. *Uncommon Ground: Toward Reinventing Nature.* New York: Norton.

Cullinan, Cormac. 2007. "Gaia's Law." In *Earthy Realism: The Meaning of Gaia,* ed. Mary Midgley. Exeter, England: Societas.

Dalby, Simon. 2002. *Environmental Security.* Minneapolis: University of Minnesota Press.

Darwin, Charles. 1884. *On the Origin of Species by Means of Natural Selection.* London: John Murray.

Davidson, Joyce, and Mick Smith. 1999. "Wittgenstein and Irigaray: Gender and Philosophy in a Language (Game) of Difference." *Hypatia: A Journal of Feminist Philosophy* 14 (2): 72–96.

Day, Richard J. F. 2005. *Gramsci Is Dead: Anarchist Currents in the Newest Social Movements.* Toronto: Pluto Press and Between the Lines.

D'Entrèves, Mauritzio Passerin. 2000. "Arendt's Theory of Judgment." In *The Cambridge Companion to Hannah Arendt,* ed. Dana Villa. Cambridge: Cambridge University Press.

de Jouvenal, Bertrand. 1957. *Sovereignty: An Inquiry into the Political Good.* Cambridge: Cambridge University Press.

Deleuze, Gilles. 1997. *Bergsonism*. New York: Zone Books.

Derrida, Jacques. 1978. *Writing and Difference*. London: Routledge.

——. 1992. "Eating Well, or the Calculation of the Subject." In *Points . . . Interviews, 1974–1994*, ed. E. Weber. Stanford, Calif.: Stanford University Press.

——. 1995a. *From Restricted to General Economy: A Hegelianism without Reserve in Writing and Difference*. London: Routledge.

——. 1995b. *The Gift of Death*. Chicago: University of Chicago Press.

——. 2008. *The Animal That Therefore I Am*. New York: Fordham University Press.

Derrida, Jacques, and Gianni Vattimo, eds. 1998. *Religion*. Stanford, Calif.: Stanford University Press.

Diehm, Christian. 2000. "Facing Nature: Levinas Beyond the Human." *Philosophy Today* 44 (1): 51–59.

Disch, Lisa Jane. 1994. *Hannah Arendt and the Limits of Philosophy*. Ithaca: Cornell University Press.

Dobson, Andrew. 1995. *Green Political Thought*. London: Routledge.

Drucker, Claudia. 1998. "Hannah Arendt on the Need for a Public Debate on Science." *Environmental Ethics* 20 (3): 305–16.

Dryzek, John. 1990. "Green Reason: Communicative Ethics for the Biosphere." *Environmental Ethics* 12: 195–210.

Dussel, Enrique. 2006. "'The Politics' by Levinas: Towards a 'Critical' Political Philosophy." In *Difficult Justice: Commentaries on Levinas and Politics*, ed. Asher Horowitz and Gad Horowitz, 78–96. Toronto: University of Toronto Press.

Eckersley, Robyn. 1992. *Environmentalism and Political Theory: Toward an Ecocentric Approach*. London: UCL Press.

——. 2004. *The Green State: Rethinking Democracy and Sovereignty*. Cambridge, Mass.: MIT Press.

Ek, Richard. 2006. "Giorgio Agamben and the Spatialities of the Camp: An Introduction." *Geografiska Annaler* 88B (4): 363–86.

Eliade, Mircea. 1987. *The Sacred and the Profane. The Nature of Religion*. San Diego: Harcourt Brace.

Elias, Norbert. 1982. *The Civilizing Process*. Oxford: Blackwell.

Esposito, Roberto. 2008. *Bios: Biopolitics and Philosophy*. Minneapolis: University of Minnesota Press.

Evernden, Neil. 1999. *The Natural Alien: Humankind and the Environment*. Toronto: University of Toronto Press.

Farias, Victor. 1989. *Heidegger and Nazism*. Philadelphia: Temple University Press.

Faun, Feral. 2000. *Feral Revolution*. London: Elephant Editions.

Featherstone, Mike, Scott Lash, and Roland Robertson, eds. 1995. *Global Modernities*. London: Sage.

Ferry, Luc. 1992. *The New Ecological Order*. Chicago: University of Chicago Press.

Flynn, Bernard. 2005. *The Philosophy of Claude Lefort: Interpreting the Political.* Evanston, Ill.: Northwestern University Press.

Foltz, Bruce V. 1995. *Inhabiting the Earth: Heidegger, Environmental Ethics, and the Metaphysics of Nature.* Atlantic Highlands, N.J.: Humanities Press.

Foucault, Michel. 1991. *Discipline and Punish: The Birth of the Prison.* Harmondsworth, England: Penguin.

———. 1994. "Omnes et singulatum: Toward a Critique of Political Reason." In *Michel Foucault: The Essential Works 3.* London: Allen Lane.

———. 2004. *Security, Territory, Population Lectures at the Collège de France 1977–1978.* Basingstoke, U.K.: Palgrave Macmillan.

———. 2008. *The Birth of Biopolitics: Lectures at the Collége de France 1978–1979.* Basingstoke, U.K.: Palgrave.

Fromm, Erich. 2001. *Fear of Freedom.* London: Routledge.

Gadamer, Hans-Georg. 1980. "Dialectic and Sophism in Plato's Seventh Letter." In *Dialogue and Dialectic: Eight Hermeneutical Studies on Plato.* New Haven, Conn.: Yale University Press.

———. 1985. *Philosophical Apprenticeships.* Cambridge Mass.: MIT Press.

———. 2000. *The Beginning of Philosophy.* New York: Continuum.

———. 2001. *Gadamer in Conversation: Reflections and Commentary.* New Haven, Conn.: Yale University Press.

Gare, Arran. 1993. *Nihilism Incorporated: European Civilization and Environmental Destruction.* Bungendore, Australia: Eco-Logical Press.

Glacken, Clarence. 1967. *Traces on the Rhodian Shore: Nature and Culture in Western Thought from Ancient Times to the End of the Eighteenth Century.* Berkeley: University of California Press.

Golley, Frank Benjamin. 1993. *A History of the Ecosystem Concept in Ecology.* New Haven, Conn.: Yale University Press.

Gourevitch, Victor. 1988. Forward. In *Critique and Crisis: Enlightenment and the Pathogenesis of Modern Society,* ed. Reinhart Koselleck. Cambridge, Mass.: MIT Press.

Gregory, Derek. 2006. "The Black Flag: Guantánamo Bay and the Space of Exception." *Geografiska Annaler* 88B (4): 405–27.

Greisch, Jean. 1991. "The Face and Reading: Immediacy and Mediation." In *Re-Reading Levinas,* ed. Robert Bernasconi and Simon Critchley. Bloomington: Indiana University Press.

Haar, Michel. 1993. *The Song of the Earth: Heidegger and the Grounds of the History of Being.* Indianapolis: Indiana University Press.

Hailes, Katherine N. 1999. *How We Became Posthuman: Virtual Bodies in Cybernetics, Literature, and Informatics.* Chicago: University of Chicago Press.

Hailwood, Simon. 2004. *How to Be a Green Liberal: Nature, Value and Liberal Philosophy.* Montreal: McGill Queen's University Press.

Hanssen, Beatrice. 1998. *Walter Benjamin's Other History: Of Stones, Animals, Human Beings and Angels.* Berkeley: University of California Press.

Haraway, Donna J. 1991. *Simians, Cyborgs, and Women: The Reinvention of Nature*. London: Free Association Books.

Harding, Stephan. 2006. *Animate Earth: Science, Intuition and Gaia*. White River Junction, Vt.: Chelsea Green Publishing.

Hardt, Michael, and Antonio Negri. 2004. *Multitude: War and Democracy in the Age of Empire*. New York: Penguin.

Hardy, Thomas. (1872) 1985. *Under the Greenwood Tree, or The Mellstock Quire: A Rural Painting of the Dutch School*. Harmondsworth, England: Penguin.

Harvey, David. 1996. *Justice, Nature and the Geography of Difference*. Oxford: Blackwell.

Hay, Peter. 2002. *A Companion to Environmental Thought*. Edinburgh: University of Edinburgh Press.

Heidegger, Martin. 1993a. "The Question concerning Technology." In *Basic Writings*. London: Routledge.

———. 1993b. "The Origin of the Work of Art." In *Basic Writings*. London: Routledge.

———. 1995. *The Fundamental Concepts of Metaphysics: World, Finitude, Solitude*. Bloomington: Indiana University Press.

———. 2001. "The Thing." In *Poetry, Language, Thought*. New York: HarperCollins.

Hendricks, Stephanie. 2005. *Divine Destruction: Wise Use, Dominion Theology, and the Making of American Environmental Policy*. Hoboken, N.J.: Melville House.

Hepburn, R. W. 1995. "The Naturalistic Fallacy." In *The Oxford Companion to Philosophy*, ed. Ted Honderich. Oxford: Oxford University Press.

Herber, Lewis. 1962. *Our Synthetic Environment*. New York: Alfred A. Knopf.

———. 1966. "Ecology and Revolutionary Thought." *Anarchy* 69: 321–40.

Hermer, Joe. 2002. *Regulating Eden: The Nature of Order in North American Parks*. Toronto: University of Toronto Press.

Hermson, Joke T. 1999. "The Impersonal and the Other: On Simone Weil (1907–43)." *European Journal of Women's Studies* 6 (2): 183–200.

Hetherington, Kevin. 1998. *Expressions of Identity: Space, Performance, Politics*. London: Sage.

Hobbes, Thomas. 1960. *Leviathan or the Matter, Forme and Power of a Commonwealth Ecclesiastical and Civil*. Oxford: Blackwell.

Honderich, Ted, ed. 1995. *The Oxford Companion to Philosophy*. Oxford: Oxford University Press.

Horowitz, Asher, and Gad Horowitz, eds. 2006. *Difficult Justice: Commentaries on Levinas and Politics*. Toronto: University of Toronto Press.

Howard, Roger. 2009. *The Arctic Gold Rush: The New Race for Tomorrow's Resources*. New York: Continuum.

Hutchens, B. C. 2005. *Jean-Luc Nancy and the Future of Philosophy*. Montreal: McGill-Queen's University Press.

Huxley, Aldous. 1971 (1932). *Brave New World*. Harmondsworth, England: Penguin.

Isaac, Jeffrey C. 1992. *Arendt, Camus, and Modern Rebellion*. New Haven, Conn.: Yale University Press.

Jensen, Derrick. 2004. "Beyond Backward and Forward: On Civilization, Sustainability and the Future." *Fifth Estate* 3 (2): 18–22.

———. 2006. *Endgame*. Vol. II, *Resistance*. New York: Seven Stories Press.

Johnston, Paul. 1989. *Wittgenstein and Moral Philosophy*. London: Routledge.

Joyner, Christopher C. 1998. *Governing the Frozen Commons: The Antarctic Regime and Environmental Protection*. Columbia: University of South Carolina Press.

Kalyvas, Andreas. 2005. "The Sovereign Weaver: Beyond the Camp." In *Politics, Metaphysics, and Death. Essays on Giorgio Agamben's Homo Sacer*, ed. Andrew Norris. Durham, N.C.: Duke University Press.

Kateb, George. 2000. "Political Action: Its Nature and Advantages." In *The Cambridge Companion to Hannah Arendt*, ed. Dana Villa. Cambridge: Cambridge University Press.

Kendall, Stuart. 2005. Editor's Introduction. In *Georges Bataille, The Cradle of Humanity: Prehistoric Art and Culture*. New York: Zone Books.

———. 2007. *Georges Bataille*. London: Reaktion Books.

Kierkegaard, Sören. 1980. *The Concept of Anxiety: A Simple Psychologically Orienting Deliberation on the Dogmatic of Hereditary Sin*. Princeton, N.J.: Princeton University Press.

Kovel, Joel. 2002. *The Enemy of Nature: The End of Capitalism or the End of the World?* Nova Scotia: Fernwood Publishing.

Kropotkin, Peter. 1913. *The Conquest of Bread*. London: Chapman and Hall.

Kuehls, Thom. 1996. *Beyond Sovereign Territory*. Minneapolis: University of Minnesota Press.

Lacan, Jacques. 1992. *Écrits: A Selection*. London: Routledge.

Latour, Bruno. 1993. *We Have Never Been Modern*. Cambridge, Mass.: Harvard University Press.

———. 2004. *Politics of Nature*. Cambridge, Mass.: Harvard University Press.

Lefebvre, Henri. 1994. *The Production of Space*. Oxford: Blackwell.

Lefort, Claude. 1986. *The Political Forms of Modern Society: Bureaucracy, Democracy, Totalitarianism*. Cambridge, Mass.: MIT Press.

———. 1988. *Democracy and Political Theory*. Oxford, England: Polity.

———. 2000. *Writing the Political Test*. Durham, N.C.: Duke University Press.

———. 2007. *Complications: Communism and the Dilemmas of Democracy*. New York: Columbia University Press.

Levi, Carlo. 2008. *Fear of Freedom*. New York: Colombia University Press.

Levinas, Emmanuel. 1985. *Ethics and Infinity: Conversations with Philippe Nemo*. Pittsburgh, Pa.: Duquesne University Press.

———. 1989. "Ethics and Politics." In *The Levinas Reader*, ed. Sean Hand. Oxford: Blackwell.

———. 1991. *Totality and Infinity.* Dordrecht: Kluwer.

———. 1996. "Substitution." In *Emmanuel Levinas: Basic Philosophical Writings,* ed. Adriaan T. Peperzak, Simon Critchley, and Robert Bernasconi. Bloomington: Indiana University Press.

———. 1998. *Entre Nous: Thinking of the Other.* London: Athlone Press.

———. 2003. *Humanism of the Other.* Urbana: University of Illinois Press.

———. 2004. *Otherwise Than Being: Or Beyond Essence.* Pittsburg, Pa.: Duquesne University Press.

Lewis-Williams, David. 2002. *The Mind in the Cave: Consciousness and the Origins of Art.* London: Thames and Hudson.

Lingis, Alphonso. 1997. "Anger." In *The Sense of Philosophy: On Jean-Luc Nancy,* eds. Darren Sheppard, Colin Sparks, and Colin Thomas. London: Routledge.

———. 2004. Foreword. In *Emmanuel Levinas Otherwise Than Being: Or Beyond Essence.* Pittsburg, Pa.: Duquesne University Press.

———. 2005. "Bare Humanity." In *Addressing Levinas,* eds. Eric Sean Nelson, Antje Kapust, and Kent Still. Evanston, Ill.: Northwestern University Press.

Lippit, Akira Mizuta. 2000. *Electric Animal: Toward a Rhetoric of Wildlife.* Minneapolis: University of Minnesota Press.

Litfin, Karen T. 1998. "The Greening of Sovereignty: An Introduction." In *The Greening of Sovereignty in World Politics,* ed. Karen T. Litfin. Cambridge, Mass.: MIT Press.

Llewelyn, John. 1991a. *The Middle Voice of Ecological Conscience: A Chiasmic Reading of Responsibility in the Neighborhood of Levinas, Heidegger and Others.* New York: St. Martin's Press.

———. 1991b. "Am I Obsessed by Bobby? (Humanism of the Other Animal)." In *Re-Reading Levinas,* eds. Robert Bernasconi and Simon Critchley. Bloomington: Indiana University Press.

———. 1991c. *The Middle Voice of Ecological Conscience: A Chiasmatic Reading of Responsibility in the Neighbourhood of Levinas, Heidegger and Others.* New York: St. Martin's Press.

Locke, John. 1988. "An Essay concerning the True Original Extent and End of Civil Government." In *Two Treatises of Government,* ed. Peter Laslett. Cambridge: Cambridge University Press.

Lovelock, James. 1987. *Gaia: A New Look at Life on Earth.* Oxford: Oxford University Press.

———. 2006. *The Revenge of Gaia.* London: Allen Lane.

Lukács, Georg. 1983. *History and Class Consciousness.* London: Merlin.

Lyons, David. 2001. *Surveillance Society: Monitoring Everyday Life.* Buckingham: Open University Press.

Marchant, Oliver. 2007. *Post-Foundational Political Thought: Political Difference in Nancy, Lefore, Badiou and Laclau.* Edinburgh: Edinburgh University Press.

Marcuse, Herbert. 1991. *One-Dimensional Man: Studies in the Ideology of Advanced Industrial Society*. London: Routledge.

Marx, Karl. 1975a. "Theses on Feuerbach VI." In *Early Writings*. Harmondsworth, England: Penguin.

———. 1975b. "Economic and Philosophical Manuscripts." In *Early Writings*. Harmondsworth, England: Penguin.

———. 1990. *Capital: Volume 1*. Harmondsworth, England: Penguin.

Masson, Jeffrey Moussaieff, and Susan McCarthy. 1995. *When Elephants Weep: The Emotional Lives of Animals*. New York: Delacorte Press.

Mathews, Freya. 1994. *The Ecological Self*. London: Routledge.

———. 2003. *For Love of Matter: A Contemporary Panpsychism*. Albany: State University of New York Press.

May, Todd. 1994. *The Political Philosophy of Poststructuralist Anarchism*. University Park: Pennsylvania State University Press.

McGregor, Gaile. 1988. *The Noble Savage in the New World Garden: Notes Toward a Syntactics of Place*. Toronto: University of Toronto Press.

McKibben, Bill. 1990. *The End of Nature*. Harmondsworth, England: Penguin.

Meadowcroft, James. 2005. "From Welfare State to Eco-State." In *The State and the Global Ecological Crisis,* eds. John Barry and Robyn Eckersely. Cambridge, Mass.: MIT Press.

Melucci, Alberto. 1989. *Nomads of the Present*. London: Hutchinson.

Merchant, Carolyn. 1989. *The Death of Nature: Women, Ecology and the Scientific Revolution*. San Francisco: Harper and Row.

Midgley, Mary. 2007. "Introduction: The Not-So-Simple Earth." In *Earthy Realism: The Meaning of Gaia*, ed. Mary Midgley. Exeter: Societas.

Miller, DeMond Shondell, Jason David Rivera, and Joel C. Yelin. 2008. "Civil Liberties: The Line Dividing Environmental Protest and Ecoterrorists." *Journal for the Study of Radicalism* 2 (1): 109–23.

Mol, Arthur P.J. 2003. *Globalization and Environmental Reform: The Ecological Modernization of the Global Economy*. Cambridge, Mass.: MIT Press.

Moore, G. E. 1922. *Principia Ethica*. Cambridge: Cambridge University Press.

Morito, Bruce. 2002. *Thinking Ecologically: Environmental Thought, Values, and Policy*. Halifax, N.S.: Fernwood.

Morris, Brian. 1996. *Ecology and Anarchism: Essays and Views on Contemporary Thought*. Worcestershire: Images.

Mouffe, Chantal. 2005. *The Return of the Political*. London: Verso.

Moyn, Samuel. 2005. *Origins of the Other: Emmanuel Levinas Between Revelation and Ethics*. Ithaca: Cornell University Press.

Murdoch, Iris. 1970. *The Sovereignty of Good*. London: Routledge and Kegan Paul.

———. 1987. Introduction. In *Sartre: Romantic Rationalist*. London: Chatto and Windus.

———. 1992. *Metaphysics as a Guide to Morals*. London: Chatto and Windus.

———. 1999. *Existentialists and Mystics*. Harmondsworth, England: Penguin.

Naess, Arne. 1972. "The Shallow and the Deep Ecology Movement." *Inquiry* 16: 95–100.

———. 1979. "Self-Realization in Mixed Communities of Humans, Bears, Sheep, and Wolves." *Inquiry* 22: 231–41.

———. 1989. *Ecology, Community and Lifestyle*. Cambridge: Cambridge University Press.

Nancy, Jean-Luc. 1991. *The Inoperative Community*. Minneapolis: University of Minnesota Press.

———. 1997. *The Sense of the World*. Minneapolis: University of Minnesota Press.

———. 2000. *Being Singular Plural*. Stanford Calif.: Stanford University Press.

———. 2007. *The Creation of the World or Globalization*. Albany: State University of New York Press.

———. 2008. "The Being-With of Being-There." *Continental Philosophy Review* 41 (5): 1–15.

Nash, Roderick. 1982. *Wilderness and the American Mind*. New Haven, Conn.: Yale University Press.

Negri, Antonio. 2008. *The Porcelain Workshop: For a New Grammar of the Political*. Los Angeles: Semiotext(e).

Ott, Hugo. 1993. *Martin Heidegger: A Political Life*. London: Fontana.

Passmore, John. 1970. *The Perfectibility of Man*. London: Duckworth.

———. 1974. *Man's Responsibility for Nature*. London: Duckworth.

Pateman, Carole. 1985. *The Problem of Political Obligation: A Critical Analysis of Liberal Theory*. Oxford, England: Polity.

Paterson, Mathew. 2000. *Understanding Global Environmental Politics: Domination, Accumulation, Resistance*. New York: Palgrave.

Peperzak, Adriaan. 1993. *To the Other: An Introduction to the Philosophy of Emmanuel Levinas*. West Lafayette, Ind.: Purdue University Press.

———. 1997. *Beyond the Philosophy of Emanuel Levinas*. Evanston, Ill.: Northwestern University Press.

———. 2005. *To the Other: An Introduction to the Philosophy of Emmanuel Levinas*. West Lafayette, Ind.: Purdue University Press.

Pepper, David. 1993. *Eco-Socialism: From Deep Ecology to Social Justice*. London: Routledge.

Perlman, Fredy. 1983. *Against His-story, Against Leviathan! An Essay*. Detroit: Black and Red.

Plato. 1963. *Plato: The Collected Dialogues*. Ed. Edith Hamilton and Huntingdon Cairns. New York: Pantheon Books.

Popper, Karl. 1969. *The Open Society and Its Enemies*. Vol. 1, *Plato*. London: Routledge & Kegan Paul.

Robertson, Roland. 1992. *Globalization, Social Theory and Global Culture*. London: Sage.

Rousseau, Jean-Jacques. 1986a. "A Discourse on a Subject Proposed by the Academy of Dijon. What Is the Origin of Inequality among Men, and Is

It Authorized by Natural Law?" In *The Social Contract and Discourses*. London: Everyman.

———. 1986b. "Essay on the Origin of Languages." In *Two Essays on the Origin of Languages*, ed. John H. Moran and Alexander Gode. Chicago: University of Chicago Press.

Routley, Val. 1981. "On Karl Marx as an Ecological Hero." *Environmental Ethics* 3: 237–44.

Sahlins, Marshall. 1972. *Stone Age Economics*. London: Tavistock.

Sandilands, Catriona. 1999. *The Good-Natured Feminist: Ecofeminism and the Quest for Democracy*. Minneapolis: University of Minnesota Press.

Sandler, Ronald, and Phaedra C. Pezzullo. 2007. *Environmental Justice and Environmentalism: The Social Justice Challenge to the Environmental Movement*. Cambridge, Mass.: MIT Press.

Saunders, Trevor J. 1984. Introduction. In *Plato: The Laws*. Harmondsworth, England: Penguin.

Schalow, Frank. 2006. *The Incarnality of Being: The Earth, Animals and the Body in Heidegger's Thought*. Albany: State University of New York Press.

Schell, Jonathan. 2002. "A Politics of Natality." *Social Research* 69 (2): 461–73.

Schivelbusch, Wolfgang. 1986. *The Railway Journey: The Industrialization of Time and Space in the 19th Century*. Leamington Spa, U.K.: Berg.

Schmeiden, Peter. 2005. "Polytheism, Monotheism and Public Space: Between Arendt and Levinas." *Critical Horizons* 6 (1): 225–37.

Schmidt, Alfred. 1971. *The Concept of Nature in Marx*. London: NLB.

Schmitt, Carl. 1985. *Political Theology*. Chicago: University of Chicago Press.

———. 2005. *Political Theology: Four Chapters on the Concept of Sovereignty*. Chicago: University of Chicago Press.

Schürmann, Reiner. 1990. *Heidegger on Being and Acting: From Principles to Anarchy*. Bloomington: Indiana University Press.

Shapin, Steven, and Simon Schaffer. 1985. *Leviathan and the Air-Pump: Hobbes, Boyle, and the Experimental Life*. Princeton, N.J.: Princeton University Press.

Shapiro, Michael J. 1999. "The Ethics of Encounter: Unreading, Unmapping the Imperium." In *Moral Spaces: Rethinking Ethics and World Politics*, ed. David Campbell and Michael J. Shapiro. Minneapolis: University of Minnesota Press.

Shrader-Frachette, Kristin. 2002. *Environmental Justice: Creating Equality, Reclaiming Democracy*. Oxford: Oxford University Press.

Simmons, William Paul. 1999. "The Third: Levinas's Theoretical Move from An-Archical Ethics to the Realm of Justice and Politics." *Philosophy and Social Criticism* 25 (6): 83–104.

Smith, Mark J. 1998. *Ecologism: Towards Ecological Citizenship*. Buckingham: Open University Press.

Smith, Mick. 1999a. "To Speak of Trees: Social Constructivism, Environmental Values and the Futures of Deep Ecology." *Environmental Ethics* 21 (4): 359–76.

———. 1999b. "Living in Integrity: A Global Ethic to Restore a Fragmented Earth." Reviewed in *Environmental Politics* 8 (4): 230–31.

———. 2001a. *An Ethics of Place: Radical Ecology, Postmodernity, and Social Theory.* Albany: State University of New York Press.

———. 2001b. "Lost for Words? Gadamer and Benjamin on the Nature of Language and the 'Language' of Nature." *Environmental Values* 120 (1): 59–75.

———. 2001c. "The Face of Nature: Environmental Ethics, Aesthetics and the Boundaries of Social Theory." *Current Sociology* 49 (1): 49–65.

———. 2001d. "Repetition and Difference: Lefebvre, Le Corbusier and Modernity's (Im)moral Landscape." *Ethics, Place and Environment* 4 (1): 31–44.

———. 2002a. "Ethical Difference(s): A Response to Maycroft on Le Corbusier and Lefebvre." *Ethics, Place and Environment* 5 (3): 260–69.

———. 2002b. "Negotiating Nature: Social Theory at Its Limits." *Environmental Politics* 11 (2): 181–86.

———. 2005a. "Hermenutics and the Culture of Birds: The Environmental Allegory of Easter Island." *Ethics, Place and Environment* 8 (1): 21–38.

———. 2005b. "Ecological Citizenship and Ethical Responsibility: Arendt, Benjamin, and Political Activism." *Environments: A Journal of Interdisciplinary Studies* 33 (3): 51–63.

———. 2005c. "Citizens, Denizens and the Res Publica: Environmental Ethics, Structures of Feeling and Political Expression." *Environmental Values* 14: 145–62.

———. 2006. "Environmental Risks and Ethical Responsibilities: Arendt, Beck and the Politics of Acting into Nature." *Environmental Ethics* 28 (3): 227–46.

———. 2007a. "Worldly (In)Difference and Ecological Ethics: Iris Murdoch and Emmanuel Levinas." *Environmental Ethics* 29 (1): 23–41.

———. 2007b. "Wild-life: Anarchy, Ecology, and Ethics." *Environmental Politics* 16 (3): 470–87.

———. 2008a. "Suspended Animation: Radical Ecology, Sovereign Powers, and Saving the (Natural) World." *Journal for the Study of Radicalism* 2 (1): 1–23.

———. 2008b. "Thinking Ecologically: A Critique." *Environments* 138 (2): 61.

———. 2010. "Epharmosis: Jean-Luc Nancy and the Political Oecology of Creation." *Environmental Ethics* 32 (4): 385–404.

Smith, Mick, and Joyce Davidson. 2008. "Civility and Etiquette." In *The Sage Companion to the City,* eds. Tim Hall, Phil Hubbard, and John Rennie Short. London: Sage.

Soper, Kate. 1995. *What Is Nature?* Oxford: Blackwell.

Spaagaren, Gert, Arthur P. Mol, and Frederick H. Buttel. 2006. *Governing Environmental Flows: Global Challenges to Social Theory.* Cambridge, Mass.: MIT Press.

Stavrakakis, Yannis. 1999. *Lacan and the Political.* London: Routledge.

Stone, Christopher D. 1988. *Earth and Other Ethics: The Case for Moral Pluralism.* New York: Harper & Row.

Sullivan, Robert J. 1985. Translator's Introduction. In *Philosophical Apprenticeships,* by Hans-Georg Gadamer. Cambridge, Mass.: MIT Press.

Taminaux, Jacques. 1997. *The Thracian Maid and the Professional Thinker: Arendt and Heidegger*. Albany: State University of New York Press.

Taylor, Charles. 1996. *Sources of the Self: The Making of Modern Identity*. Cambridge: Cambridge University Press.

———. 2007. *A Secular Age*. Cambridge, Mass.: Harvard University Press.

Taylor, Paul W. 1986. *Respect for Nature: A Theory of Environmental Ethics*. Princeton, N.J.: Princeton University Press.

Thoreau, Henry David. 1946. "Nature Essays." In *The Works of Thoreau,* ed. Henry Seidel Canby. Boston: Houghton Mifflin.

Turnbull, Colin M. 1972. *The Mountain People*. New York: Simon & Schuster.

Urry, John. 2004. *Global Complexity*. Cambridge, England: Polity.

Vail, L. M. 1972. *Heidegger and Ontological Difference*. University Park: Pennsylvania State University Press.

Vanderheiden, Steve. 2005. "Eco-terrorism or Justified Resistence: Radical Environmentalism and the 'War on Terror.'" *Politics and Society* 33 (3): 425–47.

Vaneigem, Raoul. 1994. *The Movement of the Free Spirit*. New York: Zone Books.

Vattimo, Gianni. 2004. *Nihilism and Emancipation*. New York: Columbia University Press.

Villa, Dana. 2000. "Introduction: The Development of Arendt's Political Thought." In *The Cambridge Companion to Hannah Arendt,* ed. Dana Villa. Cambridge: Cambridge University Press.

Virno, Paolo. 2004. *A Grammar of the Multitude*. Los Angeles, Calif.: Semiotext(e).

Virno, Paolo, and Michael Hardt. 1996. *Radical Thought in Italy: A Potential Politics*. Minneapolis: University of Minnesota Press.

Voltaire. 1957. *Candide*. Harmondsworth, England: Penguin.

Waldron, Jeremy. 2000. "Arendt's Constitutional Politics." In *The Cambridge Companion to Hannah Arendt,* ed. Dana Villa. Cambridge: Cambridge University Press.

Watkin, Christopher. 2007. "A Different Alterity: Jean-Luc Nancy's 'Singular Plural.'" *Paragraph* 30 (2): 50–64.

Watson, David. 1996. *Beyond Bookchin: Preface for a Future Social Ecology*. New York: Autonomedia.

———. 2004. All Isms Are Wasms. *Fifth Estate* 3 (2): 34–38.

WCED, United Nations World Commission on Environment and Development. 1987. *Our Common Future*. Oxford: Oxford University Press.

Wells, H. G. 1928. *The Open Conspiracy: Blueprints for a World Revolution*. London: Hogarth Press.

Westra, Laura. 1994. *An Environmental Proposal for Ethics: The Principle of Integrity*. Lanham, Md.: Rowman & Littlefield.

———. 1998. *Living in Integrity: A Global Ethic to Restore a Fragmented Earth*. Lanham, Md.: Rowman & Littlefield.

Whatmore, Sarah. 2002. *Hybrid Geographies*. London: Sage.

White, D. F. 2003. "Hierarchy, Domination, Nature: Considering Bookchin's Critical Social Theory." *Organization and Environment* 16 (1): 34–65.

White Jr., Lynn. 1967. "The Historic Roots of Our Ecologic Crisis." *Science* 167 (3767): 1203–7.

Whitehead, Alfred North. 1978. *Process and Reality.* New York: Free Press.

Whiteside, Kerry. 1994. "Hannah Arendt and Ecological Politics." *Environmental Ethics* 16 (4): 339–58.

Widdows, Heather. 2005. *The Moral Vision of Iris Murdoch.* Aldershot, U.K.: Ashgate.

Wilkinson, Loren, ed. 1991. *Earthkeeping in the 90s: Stewardship of Creation.* Grand Rapids, Mich.: William B. Eerdmans.

William, Michael. 2001. "Why I Am a Primitivist." *Anarchy: A Journal of Desire Armed* 51: 39, 59.

Wissenburg, Marcel. 1998. *Green Liberalism: The Free and Green Society.* London: UCL Press.

Wittgenstein, Ludwig. 1988. *Philosophical Investigations.* Oxford: Blackwell.

———. 1990. *Tractatus Logico-Philosophicus.* London: Routledge.

Wood, David. 1999. "Comment ne pas manger—Deconstruction and Humanism." In *Animal Others: On Ethics, Ontology, and Animal Life,* ed. H. Peter Steeves, 15–35. Albany: State University of New York Press.

———. 2005. "Some Questions to My Levinasian Friends." In *Addressing Levinas,* eds. Eric Sean Nelson, Antje Kapust, and Kent Still. Evanston, Ill.: Northwestern University Press.

Woodcock, George. 1975. *Anarchism.* Harmondsworth, England: Penguin Books.

World Wildlife Fund. 2008. *A New Sea: The Need for a Regional Agreement on the Management and Conservation of the Arctic Marine Environment.*

Wyschogrod, Edith. 2000. *Emmanuel Levinas: The Problem of Ethical Metaphysics.* New York: Fordham University Press.

Young-Bruehl, Elisabeth. 2006. *Why Arendt Matters.* New Haven, Conn.: Yale University Press.

Zerzan, John. 1994. *Future Primitive and Other Essays.* New York: Autonomedia.

———. 1999. *Elements of Refusal.* Columbia, Mo.: Columbia Alternative Library Press.

———, ed. 2005. *Against Civilization: Readings and Reflections.* Los Angeles: Feral House.

Zimmerman, Michael. 1994. *Contesting Earth's Future: Radical Ecology and Postmodernity.* Berkeley: University of California Press.

———. 1995. "Heidegger, Buddhism, and Deep Ecology." In *The Cambridge Companion to Heidegger,* ed. Charles Guignon. Cambridge: Cambridge University Press.

Žižek, Slavoj. 2001. *Did Somebody Say Totalitarianism? Five Interventions in the (Mis)use of a Notion.* London: Verso.

———. 2002. *Welcome to the Desert of the Real.* London: Verso.

———. 2008. *In Defence of Lost Causes.* London: Verso.

INDEX

MICK SMITH is associate professor and Queen's National Scholar in the philosophy department and the School of Environmental Studies at Queen's University, Kingston, Ontario. He is author of *An Ethics of Place: Radical Ecology, Postmodernity, and Social Theory.*